Exploring
Human Geography
with Maps

Margaret W. Pearce
Western Michigan University

W. H. Freeman and Company
New York

ISBN-13: 978-0-7167-4917-2
ISBN-10: 0-7167-4917-3

Maps and figures by Margaret W. Pearce unless otherwise indicated.

Printed in the United States of America

Fourth printing

W. H. Freeman and Company
41 Madison Avenue
New York, NY 10010
Houndmills, Basingstoke
RG21 6XS, England
www.whfreeman.com

Contents

Introduction

This book of map exercises and activities introduces you, the geography student, to the diverse world of maps as a fundamental tool for exploring and presenting ideas in human geography. The book takes a thematic approach to the subject, exploring one geographic theme per chapter with exercises in map interpretation and construction.

The purpose of the exercises is to reinforce rather than repeat information from your textbook and class. Each map exercise applies fundamentals of human geography while also introducing map construction and interpretation skills.

Structure of Each Chapter

Each chapter begins with a vocabulary box listing the human geography terms you need to know to do the map exercises. This vocabulary is not defined in the exercises, so it will help to take a moment to review their meanings from your class and reading notes before starting the exercises.

Below the list of applied vocabulary is a list of new vocabulary applied in that chapter. You will see that there is more new vocabulary to learn in the earlier chapters; these taper off as you gradually master the basics.

Because the exercise chapters are intended to be completed in chronological order, later chapters build on the terminology and mapping skills of the earlier chapters. If you choose to dip into the book for short sections, or approach the topics in a different order, referring to these vocabulary boxes will help you locate unfamiliar concepts or Web sites as you proceed.

After the vocabulary boxes, each chapter (except the first and the last) proceeds with three map exercises exploring some of the themes from class. There are three modes of exercises, reflecting the three modes of the map as a tool for geographers. The mode of an exercise is shown at the beginning of the exercise by an icon so that you know which type of map exercise it is. (Some exercises explore more than one mode of the map and have two icons.) The three modes are:

THE LANGUAGE OF MAPS

This mode introduces the fundamentals of map use and interpretation, such as scale, projections, data classification, and remote sensing, in human geographical contexts.

VISUAL EXPLORATIONS

These exercises focus on using maps for exploring different representations of spatial data. Most of the exercises in this mode use interactive Web sites for data exploration. In each case, the Web URL is given in the text. The exercises can also be accessed through the W. H. Freeman Web site at **www.whfreeman.com/ jordan**. To complete these exercises, you only need access to an Internet browser. Some of the URLs utilize java plug-ins and/or Adobe Acrobat Reader. Apart from these tools, however, no special software is needed.

OTHER WAYS OF MAPPING

These exercises introduce maps approaching the theme from the perspective of another culture or period in history. The exploration and depiction of the human dimension through maps is not an invention of modern Western society, and there is much that can be learned by looking at how others have approached it.

For most of the exercises in the three modes, you will be interpreting maps and writing the answers on a separate piece of paper. In some instances, a map in the book is intended to be photocopied as part of the assignment, or you will be asked to print a page from a Web site. The book was designed to stay intact this way so that you can always refer back to maps and vocabulary in completed exercises, and so you can make use of the index.

In addition to access to the Internet and a photocopier, you will occasionally need to use a calculator for the exercises, and one exercise calls for you to use a compass to draw circle radii. There is no other special equipment required.

Why Only Maps?

Human geographers depend on all kinds of tools and materials beyond writing to do their work because the work itself is diverse. They need tools for interpreting human landscapes, for collecting and analyzing geographical information, and for presenting their findings to others. Maps permeate each of these stages of doing human geography, from initial source materials to exploration and analysis to communication of findings.

Rather than explore the spectrum of different methods and sources available to geographers, the focus here is on the map, in all of its diverse applications. It is hoped that you will take away from these exercises a vision of mapping as rich, multifaceted, fluid, and fundamental to your awareness of human geography.

CHAPTER

1

Maps for Human Geography

**Vocabulary applied in
this chapter**
spatiality
map
culture
material culture

New vocabulary
definitions and functions
  of the map
map symbols:
  mimetic and abstract

✳ 🌱 1.1
Maps and Culture

The task of the geographer is to investigate **spatiality.** But spatiality is an elusive concept to express. Often when we attempt to explain it in words, spatiality slips away. Words, whether spoken or written, are a linear mode of explaining, and as such are limited in their communication of spatial concepts.

Alternatively, if we attempt to explain a place with a picture, as in a drawing or a photograph, another limited dimension of spatiality is presented. Pictures capture one frame, one piece of space, but spatiality itself is fragmented by these frames. How does this frame relate to another frame? Where are the interconnections?

To overcome these limitations, geographers often use words and pictures together to approximate spatiality. But there is another way to both explore and express this spatiality — mapping.

As a visual medium, the **map** is one of the strongest tools for communicating spatiality. The characteristics of the map, especially its blending of scientific and artistic aspects, render it particularly useful for geography. Indeed, many people view the map as the most important tool that a geographer can learn to use. This chapter explores three aspects of the map that make it so useful.

Maps Show Location

Maps provide locational information — the first clues to understanding place. Maps show us the nature of the connections between those locations, how they are joined or separated. Connections allow us to stand back and see "the big picture" that may be invisible to us as we are standing in the landscape. For example, we may observe the distribution of phenomena over long distances, and the density and sparsity of things observable and physical or abstract and invisible.

Study the map on the facing page. This is a detail of the Door Peninsula region from the 1996 map, "Cultural Map of Wisconsin." The map portrays the cultural geography of the state through detailed geographical information and insets for the major cities. Each point symbol refers to a type of cultural feature by shape and color (the numbers refer the reader to an accompanying index booklet for information). Each symbol is defined in the map's legend.

Question 1: Reflect on the characteristics of culture from your textbook. What is culture? Make a list of the essential elements that comprise culture.

Question 2: How do you think the makers of the map define culture? What types of elements are included in their definition?

Question 3: Compare the cultural criteria from Questions 1 and 2. What elements are similar? What elements are different?

Question 4: Which elements of culture do you think would be most difficult to depict on a map?

In his classic work *Things Maps Don't Tell Us*, Armin Lobeck wrote that in order to understand the physical landscape, geographers must be able to interpret the physical processes behind the symbols on the map. The map presents the results but omits the process, and only through learning how to closely read the map can a geographer access the missing dimensions of the physical world.

OG7	College or University
DA22	Festival
IA10	Hiking Trail
IK2	Historic Community
TA3	Historic Site
RA15	Lighthouse
CLS	Museum or Tour
FT7	Park or Forest
OG11	Rustic Road
SA32	Writer

The **White Rapids of the Menominee River** was an ancestral home of the region's Native Americans. In the 1920s the Indian people were relocated onto the Menominee Reservation.

Before Euro-Americans renamed them, Washington and surrounding islands were called the Huron or Potawatomi Islands by local Indian peoples. These earliest inhabitants were attracted by the islands' excellent fisheries and natural protection, qualities that drew Norwegians, Danes, and Icelanders (rare among Wisconsin's immigrants) here in the mid-19th century. Tourism has long since replaced agriculture on **Washington Island**, but the descendants of European immigrants have persisted, celebrating their heritage every August with the annual Scandinavian Festival.

Archeological evidence along the **Rock Island State Park** suggest presence of the Potawatomi, Fox, and Menominee. The park is also home the Thordarson Estate Historic Distric a Chicago electrical inventor and Icelandic descent. Its design is ba parliament, and its massive oak carved by an Icelandic crafts

The Clearing in Ellison Bay was originally the summer home of eminent landscape architect Jens Jensen and served as his school from 1936 until his death in 1951. Today the school continues to provide a unique learning environment committed to the Danish immigrant's vision of improving the human condition through a closer relationship with nature.

In the days before modern navigational treacherous crossin **Morts Passage** (Dea source of great fear and 1889, one light recorded two shipwr The strait may have the loss of an Ind caught in a sudd

The **Crivitz Area Museum**, opened in 1990, depicts the early life of the Crivitz region, focusing on its heritage and people—many of whom were of Polish descent—with exhibits on American Indians, lumbering, railroads, pioneers, and the early resort industry. MR2

Capitalizing on what lumberman Isaac Stephenson called the "bountiful providence" of the Menominee River's 4,000-square-mile watershed and relying on French-Canadian, German, and Swedish labor, **Marinette** and sister city Menominee in Michigan became the primary lumber-milling cities in the region by the 1880s. As the great northern pine forests were exhausted, their economies diversified to paper, chemicals, castings, twine and netting, and automotive parts.

In *Old Peninsula Days*, Hjalmar Holand noted that not many years ago in **Fish Creek** "nearly every man in the village was a fisherman, and north and south of the village the shore was lined with fishermen's homes and nets." By the turn of the century, however, the community had already begun the transition to tourism, and by 1917 summer hotels served 300 guests, the largest number in Door County.

Norbert Blei composed his three "Door" books just outside Ellison Bay. DR6

In 1853, after moves from Milwaukee and Howard and some initial disagreement ov issue of collective versus individual prop ownership, a group of Norwegian Morav established the religious colony they cal **Ephraim** (meaning 'the very fruitful'). Too number of the original buildings of th Moravian colony may be seen.

The **Lower Sugar Bush German settlement** southwest of Peshtigo dates to the mid-19th century. Many of the pioneer-hewn log buildings are still in use. MR4

Largely a result of indiscriminate lumbering and the carelessness of farmers, rail workers, and lumberjacks, the **Great Peshtigo Fire** of October 8, 1871, engulfed an estimated 2,400 square miles of land on both sides of Green Bay. It left 1,500 dead—800 in Peshtigo alone—and was the worst disaster of its type in the history of the United States, occurring on the same day as the more famous but less lethal Chicago fire. MR7

Combined as a single Natural Landmark in 19 **Sanctuary**, Toft's Point, and the Mud Lake **Area** just east of Bailey's Harbor comprise a ser ridges and swales of ancient boreal forest vegetation. Without the determined effort of the Women's Club in pushing for the preservation of Sanctuary in the Great Depression, the land would have been converted to a trailer park

Oconto originally was the site of a Menominee Indian village and a short-lived Jesuit mission founded by Father Claude Allouez in 1669. Later, Oconto became an important processing and shipping point for the late 19th-century log harvest along the Oconto River. The wealth generated by the lumbering era is evident in its 19th-century buildings.

Honoring the Peninsula's first known inhabitants, the Potawatomi and Menominee Indians, a 30-foot totem pole stands on the **Peninsula State Park** Golf Course west of Ephraim. Chief Simon Kahquados of the Potawatomi unveiled the original monument in 1927 (today's pole is a more detailed reproduction of the original) and, upon his death four years later, was buried a few feet from the pole under a glacial boulder. DR27

French fur traders recognized **Green Bay's** locational advantages early on, and by 1680 La Baye, as it was known to the French, emerged as an outpost for their fur-trade and missionary activities. After the Americans took possession in 1816 following 50 years of British control, Green Bay slowly became a lumber port and immigrant outfitting center for northeastern Wisconsin. Harbor improvements, the Fox River canal system, and new rail connections helped the city attract industrial development that beckoned thousands of European immigrants. Today, manufacturers of paper and paper products, along with food processors and packers, are the city's largest employers.

The communities of Brussels, Namur, Rosiere, and Dyckesville make up the core of the largest **Belgian rural settlement** in the United States. Groups of Walloon (French-speaking) Belgian immigrants began arriving in 1853 and were later joined by Flemings as well. The area, designated a National Historic Landmark in 1990, has been declared Wisconsin's first rural National Historic District. DR7

Sturgeon Bay's strategic location at a portage between the head of the bay and Lake Michigan was well known to Native Americans and early Euro-Americans. The Sturgeon Bay Ship Canal, constructed by the early 1880s with considerable federal funding, linked Green Bay with Lake Michigan and attracted important maritime industries. Sturgeon Bay is also the shipping, packing, and processing center for locally grown cherries, a nationally recognized crop since the turn of the century.

Styling itself the 'Trout and Salmon Capital of Wisconsin,' **Algoma** currently boasts the state record chinook salmon catch from Lake Michigan. In its infancy, however, it rivalled neighboring Kewaunee for the region's lumber trade. By the late 19th century the town had become a marketplace and supply center for the newly created dairy and grain farms of the northern cutover lands.

The **Von Stiehl Winery**, housed in the former Ahnapee Brewery in Algoma, was built by Bohemian immigrants in 1866. Specializing in traditional fruit wines using locally grown cherries, the winery has won several national and international awards. KW8

Long the home of Potowatomi Indians, **Kewaunee** became the site of intense land speculation and Euro-American settlement in 1836 following reports of gold found in nearby swampy lands. The market crashed the very next year, but revival came in the 1840s when Kewaunee began to grow as a lumbering town. As the northern forests declined, Kewaunee utilized its fine harbor to become an east-west transshipment point and a manufacturing town.

Public criticism of severe pollution and dying fish associated with paper mill and other industrial waste in the **Lower Fox River** dates to the 1920s. However, federal legislation in 1965 and 1972 led to what is now a world-famous clean-up program

Since their initial settlement in 1854, **Bohemians** have been one of the most important ethnic groups in Kewaunee County. The hamlets of Strangelville, Krok, Pilsen, and

The first lighthouse at **Rawley Point** in Point Beach State Forest dates to 1853. The present steel tower was built in 1894 and is named for Peter Rawley, believed to be the area's first Euro-American settler. At least 10 shipwrecks occurred here in the 1850s

White Rapids Dam

pike River

Amberg

MR1

Wausaukee

Lake Noquebay

Crivitz

MR3

s Flowage River was portant quartz

Marinette

MR5

Peshtigo River

Pound

180

64

141

Coleman

Peshtigo

MR7 MR4

MR4

41

Lena

Oconto

OC1 OC3 OC4
OC6 OC7

Pensaukee

Gilson
MR9

GREEN BAY

DOOR

Chambers Island

DR9

DR16
DR17 DR27

Fish Creek

DR25 DR39

DR26

Egg Harbor

DR11

DR14

DR10 DR6

Ellison Bay

DR22

DR12 DR33

Sister Bay

Ephraim

Bailey's Harbor

DR30 DR4

DR5

DR3

DR8

42

57

DR42

DR15

DR29

Sturgeon Bay

DR24

DR34

DR13 DR14 DR20 DR21 DR37

DR36

DR19

DR31
DR32 DR38

Rock Island

DR41 DR18

Washington Island

Porte des Morts Passe

Gills Rock

DR28

Kangaroo Lake

42

141
41

ams

BR1

BR1

BR5

See Inset on Reverse

Green Bay

Allouez

Bellevue

De Pere

43

29

54

163

Luxemburg

Pilsen
KW2

KW7

Stangelville

42

Denmark

BROWN

KEWAUNEE

DR7 Brussels

DR40

57

DR1

Forestville

42

Algoma

KW4 KW8

KW5

Kewaunee
KW1 KW3 KW6

Detail (above) and legend (left) from "Cultural Map of Wisconsin" by David Woodward, Robert Ostergren, Onno Brouwer, Steven Hoelscher, and Joshua G. Hane. © 1996 University of Wisconsin Press. Used with permission of author and publisher.

The same can be said for culture. Often, the map depicts the locations of the remnants or artifacts of cultural processes, and the geographer must learn how to read closely to find the missing dimensions of cultural geography.

Throughout this book, we will try to push the limits between what can be *seen* in a map and what can be *gleaned from what is seen*, in order to become more adept users and critical viewers of maps.

Part of reading closely begins with looking at the overall pattern of elements shown in the map, and in the connections between elements in the pattern.

Question 5: Look again at the elements depicted in the Door Peninsula map, focusing this time on connections between locations. Can you glean any information about cultural processes from these connections?

Maps are Exploration Tools

In addition to showing location, maps are also **tools for exploring a geographical idea or problem.** Geographical ideas are often explored through discussion and debate, whether in written or spoken words. But because they are spatial, geographical ideas can also be explored visually, through graphic representation and rerepresentation, until they are solved.

An example of this type of exploratory mapping can be seen in the maps collected by Hugh Brody in

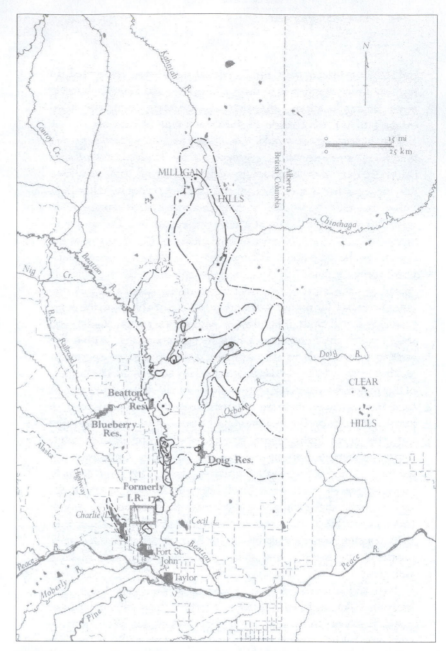

"Doig River Reserve: One Hunter's Land-Use Biography," from *Maps and Dreams: Indians and the British Columbia Frontier* by Hugh Brody. Prospect Heights, Ill.: Waveland Press, 1997. Used with permission of the author. Map by Karen Ewing.

British Columbia in 1978. Brody's task was to map the hunting areas of native people in northern British Columbia, as part of a federal government effort to anticipate the effects of the siting and construction of the Alaska Oil and Gas Pipeline through the region.

Brody began by asking individuals to mark on maps the bound-

aries of the lands they used for hunting, fishing, and berry-picking, creating cartographic hunting biographies for each person, as in the map above. He then compiled all the maps for each activity into one map, showing all berry-picking territory, all hunting territory, and all fishing territory, as overlapping boundaries, as shown in the four maps on page 7.

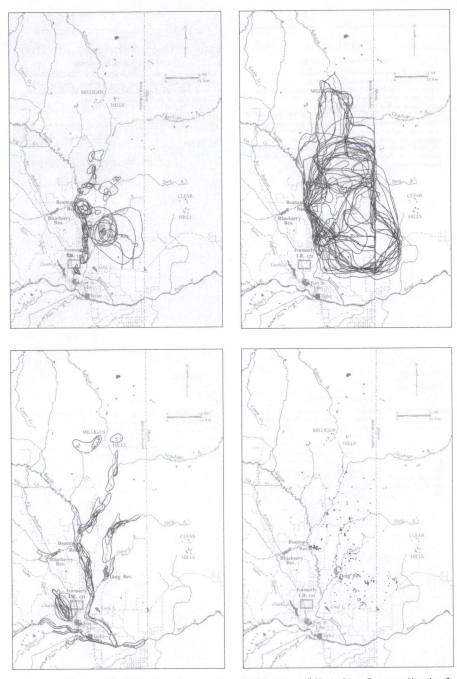

Clockwise from top left: "Doig River Reserve: Berry Picking Areas," "Doig River Reserve: Hunting," "Doig River Reserve: Camping Sites," "Doig River Reserve: Fishing Areas," from *Maps and Dreams: Indians and the British Columbia Frontier* by Hugh Brody. Prospect Heights, Ill.: Waveland Press, 1997. Used with permission of the author. Maps by Karen Ewing.

Question 6: What are the factors defining culture in this region? What aspect of those factors is depicted in the map?

Question 7: Compare the map "biography" with the activity maps. What do you think is the difference between the spatiality of culture as it is manifested in the activities of an individual, versus culture as it is manifested in the activities of a community?

Later, Brody compiled the activity maps into a base map showing the outline of the estimated total area required for sustenance by each of the reserves, shown below.

Question 8: What kind of cultural information was ultimately lost in the final, compiled map?

The map compilations technically answered the question that Brody had been hired to find, the visual picture of hunting, fishing, and berry-picking boundaries. In the process, however, he also began to realize the limitations of such maps to communicate other cultural forces in the region, such as the tension between native and nonnative perceptions of the land, and the relationship between people, animals, and seasons. Brody's recollection of his experience in British Columbia is a classic study of both the use of maps to explore a cultural geographical problem, as well as the limitations of this type of map as a means to convey cultural connections to the land.

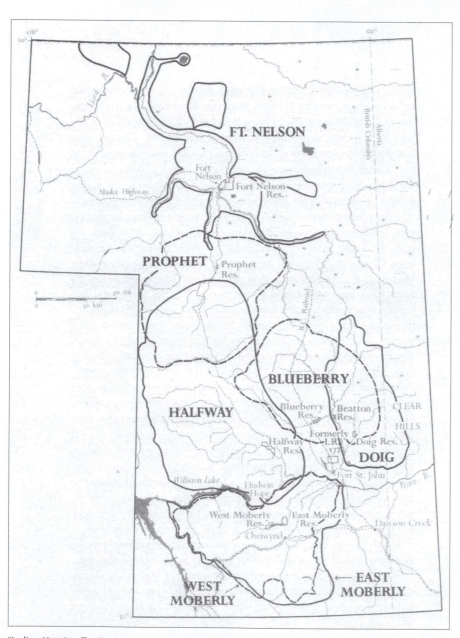

"Indian Hunting Territories in Northeast British Columbia," from *Maps and Dreams: Indians and the British Columbia Frontier* by Hugh Brody. Prospect Heights, Ill.: Waveland Press, 1997. Used with permission of the author. Map by Karen Ewing.

Maps are Material Culture

Finally, maps are also interesting to geographers because they are themselves **material culture** or **cultural evidence**. Maps locate and explore culture, and at the same time are a part of culture. As material culture, maps are primary sources that can graphically portray the perceptions, priorities, conventions, and aesthetics of a people.

A map does not merely record a landscape, it records the mapmaker's perception of that landscape, and the mapmaker's effort to shape our understanding. The mapmaker is shaped by culture. As a result, the map is not only a means of interpreting the cultural landscape, but also, as an artifact, is itself a part of that cultural landscape.

Because cultures differ greatly from one another, the look of the maps that each culture produces is also marked by great contrast. The differences in the way cultures perceive and represent the world are reflected in every element of the map: the amount of detail shown, the direction in which the map is oriented, the proportion of words versus graphics in the map, even the way the map symbols are conceived and drawn.

For example, we can find many cultural clues by studying the legend of a map. A map's legend often defines those features considered most significant by the mapmaker or institution behind the map. The legend above is

"Symbols for Village Maps," from *Maya Atlas: The Struggle to Preserve Maya Land in Southern Belize*, © 1997 North Atlantic Books. Used with permission of the publisher.

from the 1997 *Maya Atlas*, a collection of maps drawn by the Maya People of Southern Belize, depicting thirty-six Mayan communities. The goal of the atlas was to clearly represent Mayan lands without using the maps of other governments as a base, in order to establish a legal basis for indigenous land rights and land claims in the region.

Question 9: Based on the symbols defined in the legend, what assumptions would you make about daily life in Southern Belize? How do people make a living? What do they do for entertainment?

Question 10: Based on this legend, what assumptions would you make about the physical landscape?

Map symbols can vary dramatically from culture to culture. Every culture has a different sense of how **mimetic** a map symbol should be, that is, to what degree the symbol should resemble the object it represents, or by contrast how **abstract** a symbol should be. (For example, to symbolize a campground on a map, a tent would be a mimetic symbol, and a dot or square would be an abstract symbol.) The level of abstraction may vary within a culture, too, because cultures develop different types of maps to serve different needs in their societies.

The illustration on the right depicts a selection of symbols from Mesoamerican (both Mixtec and Aztec) cartography. On page 11, a selection of symbols from the U.S. Geological Survey topographical map series is shown for comparison. The two examples are not "map symbols" in quite the same sense of the term. The Aztecs used a pictographic, not an alphanumeric, writing system, so unlike Western map symbols, their map symbols are both the symbol and place-name combined. But we can still make comparisons based on the functions of the map symbols.

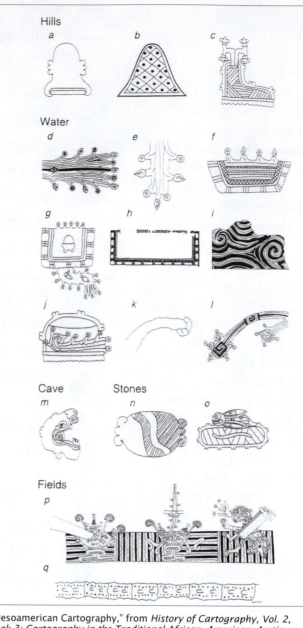

"Mesoamerican Cartography," from *History of Cartography, Vol. 2, Book 3: Cartography in the Traditional African, American, Arctic, Australian, & Pacific Societies,* © 1998 University of Chicago Press. Used with permission of the publisher.

Question 11: Compare the symbols from the Maya Atlas on page 9 to the Mesoamerican symbols, above, and the U.S. Geological Survey symbols. Each symbol set comes from a different culture's cartographic "language." Which symbol set is the most abstract? Which is the most mimetic?

Question 12: Are there wide differences in abstract and mimetic symbols in the same symbol set? Why do you think this is?

Question 13: All three cultures' symbol sets have a symbol for "cave." How do they compare?

Question 14: Does a comparison of the symbol for "cave" across three cultures give any clues to those cultures' different perceptions of the natural world? Why or why not?

RIVERS, LAKES, AND CANALS

Intermittent stream	
Intermittent river	
Disappearing stream	
Perennial stream	
Perennial river	
Small falls; small rapids	
Large falls; large rapids	
Masonry dam	
Dam with lock	
Dam carrying road	
Perennial lake; Intermittent lake or pond	
Dry lake	
Narrow wash	
Wide wash	
Canal, flume, or aqueduct with lock	
Elevated aqueduct, flume, or conduit	
Aqueduct tunnel	
Well or spring; spring or seep	

GLACIERS AND PERMANENT SNOWFIELDS

Contours and limits	
Form lines	

COASTAL FEATURES

Foreshore flat	
Rock or coral reef	
Rock bare or awash	
Group of rocks bare or awash	
Exposed wreck	
Depth curve; sounding	
Breakwater, pier, jetty, or wharf	
Seawall	

VEGETATION

Woods	
Scrub	
Orchard	
Vineyard	
Mangrove	

SURFACE FEATURES

Levee	
Sand or mud area, dunes, or shifting sand	
Intricate surface area	
Gravel beach or glacial moraine	
Tailings pond	

MINES AND CAVES

Quarry or open pit mine	
Gravel, sand, clay, or borrow pit	
Mine tunnel or cave entrance	
Prospect; mine shaft	
Mine dump	
Tailings	

BUILDINGS AND RELATED FEATURES

Building	
School; church	
Built-up Area	
Racetrack	
Airport	
Landing strip	
Well (other than water); windmill	
Tanks	
Covered reservoir	
Gaging station	
Landmark object (feature as labeled)	
Campground; picnic area	
Cemetery: small; large	

Detail from "Topographic Map Symbols," U.S. Geological Survey, National Mapping Information, http://mac.usgs.gov/mac/isb/pubs/booklets/symbols/index.html.

Sources and Suggested Reading

Maps and Culture

Brody, Hugh. *Maps and Dreams: Indians and the British Columbia Frontier.* Prospect Heights, Ill.: Waveland Press, 1997.

Godlewska, Anne. "The Idea of the Map," in Susan Hanson (ed.). *Ten Geographic Ideas that Changed the World.* New Brunswick, N.J.: Rutgers University Press, 1997, pp. 17–39.

Lobeck, Armin K. *Things Maps Don't Tell Us.* Chicago: University of Chicago Press, 1960.

MacEachren, Alan M. *Some Truth with Maps: A Primer on Symbolization and Design.* Washington, D.C.: Association of American Geographers, 1994.

Mundy, Barbara E. "Mesoamerican Cartography," in David Woodward and G. Malcolm Lewis (eds.). *History of Cartography, Volume 2, Book 3: Cartography in the Traditional African, American, Arctic, Australian, and Pacific Societies.* Chicago: University of Chicago Press, 1998.

Toledo Maya Cultural Council. *Maya Atlas: The Struggle to Preserve Maya Land in Southern Belize.* Berkeley, Calif.: North Atlantic Books, 1997.

Woodward, David. "Cultural Map of Wisconsin." [map] 1:500,000. Madison, Wis.: University of Wisconsin Press, 1996.

2

Folk Culture / Popular Culture

Vocabulary applied in this chapter
folk landscape
relict landscape
popular landscape

New vocabulary
map scale:
   large-scale map
   small-scale map
representative fraction (RF)
map perspective:
  planar, oblique, profile
popular cartography
folk cartography

✷ 2.1
Scale and
Cultural Evidence

One of the most fundamental pieces of information that a geographer needs to know about a map, to decide if it will make a good source, is **map scale.** Map scale, because it is inextricably tied to detail, often determines the type of cultural information that can be extracted from the map.

A **large-scale** map covers a small surface area in high detail. A typical large-scale map might include local roads, building footprints, vegetation, or elevation.

A **small-scale** map is the opposite: it covers a large surface area in low detail. A typical small-scale map would show major highways as lines, towns and cities as points, or whole continents or hemispheres.

Scale can be represented on a map with a bar for measuring lengths, or as a statement, as in: "one inch equals five miles."

Scale can also be represented by a **representative fraction,** or **RF,** which tells you the ratio of the relationship of the map to earth. In other words, an RF of 1:10,000 means that 1 length of anything on the map equals 10,000 of those same lengths on the earth: 1 inch on the map would equal 10,000 inches on the earth, 1 millimeter on the map would equal 10,000 millimeters

on the earth, and 1 handlength on the map would equal 10,000 handlengths on the earth.

In the language of the RF, anything that is 1:25,000 scale or larger is considered "large scale." Anything 1:250,000 scale or smaller is considered "small scale." (And everything in the middle is "medium.")

The scale of a map, whether it is large or small, can be a good indicator of the type of geographical information that you will find in the map. Consider two examples from the Ordnance Survey map series from the United Kingdom. The map below is 1:63,360, which seems to be an odd RF, until you translate it to the equiv-

alent verbal statement: "one inch equals one mile." A map of the same area on the opposite page is 1:25,000. Between 1:63,360 and 1:25,000, the difference in the amount of geographical detail that can be shown is great.

Question 1: Study the geographical features of the 1:25,000 map. What kinds of transportation routes are depicted? What kinds of features are named? Are there any indicators of the type of economic activity in this region?

Question 2: Compare your reading to the smaller-scale 1:63,000 map below. What features are preserved in both maps? What information is lost in the smaller-scale map?

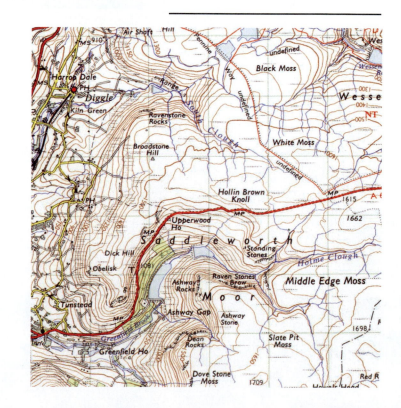

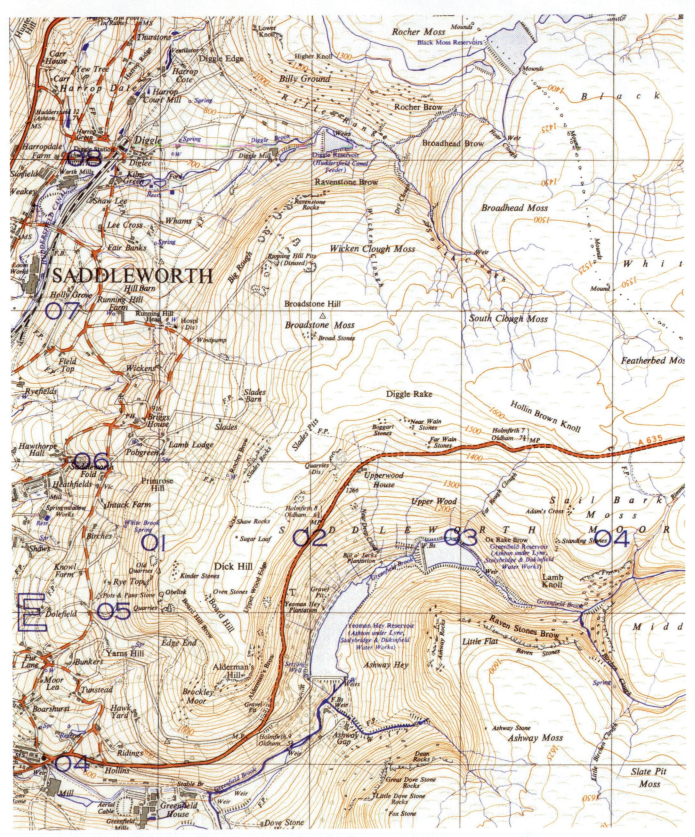

Left: Detail from "Huddersfield," 1961. Above: detail from "Saddleworth," 1957. Both images reproduced from Ordnance Survey mapping on behalf of The Controller of Her Majesty's Stationery Office, © Crown Copyright. License Number MC 100037914.

Question 3: How would your impression of this region be different if you had only the 1:63,360 map to work with?

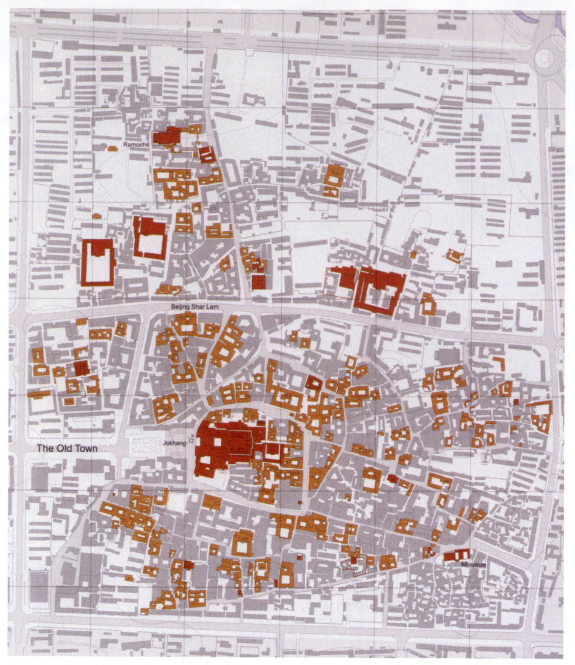

Detail of "General map of the LHCA project, 1999," from *The Lhasa Atlas: Traditional Tibetan Architecture and Townscape*, © 2001 Knud Larsen and Amund Sinding Larsen. Used with permission of the author and the publisher.

Finding Parallel Worlds in the Large-Scale Map

Large-scale maps are a useful way to look for examples of the parallel worlds of the **folk** and **popular landscapes** because large-scale maps have more potential to show material culture than do small-scale maps.

In this map of the ancient city of Lhasa, Tibet, we can explore these parallel worlds. Looking at the city at an RF of 1:7,500,

the patterns of city streets and buildings are clearly visible. Modern architecture is depicted in gray. Traditional architecture is depicted in dark orange, signifying religious buildings, and light orange, signifying secular buildings. Tibetan urban form is differentiated from modern city planning by certain characteristics, including orientation of buildings to the south; use of circular structure to emphasize a center focal point (rather than the West-

ern structural tradition of a linear line of sight terminating at a focal point); and curved rather than straight paths and roads.

The map was compiled by the Lhasa Historical City Atlas Project as part of an effort both to document the rapidly changing Lhasa architectural pattern from traditional Tibetan to modern Chinese, and to establish a basis for Tibetan architectural preservation. In Lhasa, as in many of

Clockwise from top left: "New Cinema Building," "New Market Hall," and "Tsonak Lam," by Knud Larsen, © 2001. From *The Lhasa Atlas: Traditional Tibetan Architecture and Townscape* by Knud Larsen and Amund Sinding Larsen. Used with permission of the author and the publisher.

the world's old cities, the ancient vernacular architecture is at risk of being replaced with modern buildings. The Lhasa photos on this page show the difference between the modern architecture of China (top left) and the vernacular architecture of Tibet (below left). The photo above, right shows how the two types of built landscapes parallel each other at a marketplace.

A look at the map on the opposite page immediately shows the parallel traditional (orange) and modern (grey) parts of the city. But even without the benefits of color coding, closer study of the map reveals many indicators of the parallel worlds.

Question 4: What differences in the shapes and sizes of the streets indicate differences between vernacular and modern?

Question 5: Are there differences in the orientation of streets and buildings, which would also indicate vernacular or modern landscapes? Explain why you think they are similar or different.

Question 6: The photographs on this page illustrate the differences in the two architectural styles. Can you also see differences between the two styles in the map? If so, how do these differences compare to what is shown in the photographs?

Much of the Lhasa folk landscape has already been covered over by modernity. For example, many of the ancient concentric pilgrimage routes, called koras, are now part of the **relict landscape**. The koras are the old circular paths that form concentric circles around the major temples.

Question 7: Where in the map can you find the relic patterns of the koras? What aspects of the built environment reveal the locations of these earlier paths?

✷ ◑ 2.2
Putting It All in Perspective

Like map scale, **map perspective** is another characteristic that determines the type of geographical information that can be portrayed in a map. Perspective is the term that cartographers use to describe the angle from which a map or image is viewed. All maps present the landscape from a particular angle for the viewer, though as viewers we may not always be aware of it.

A map that presents the landscape as if we are looking straight down on it from above is called a **planar** view. This is the conventional perspective view of a map, as shown in the early U.S. Geological Survey map of St. Paul, Minnesota on the lower half of the opposite page.

Because the U.S. Geological Survey publishes several editions of the same map over time, to account for changes in both the physical and built landscapes, geographers often draw on several editions of the same map sheet to visualize geographical change in a given area.

A map that presents the landscape as if we are looking down upon it from above, at a slight angle, is called an **oblique** or **bird's-eye view.** The image at the top of the opposite page depicts St. Paul from an oblique perspective.

Question 1: What geographical information is unique to the oblique perspective? To the planar perspective?

Question 2: Are there any geographical elements that are downplayed or hidden by oblique perspective? Explain.

Question 3: How does the planar perspective map influence your perception of the elements noted in Question 2? In other words, does the planar perspective provide a more balanced view of the city, or does it emphasize and deemphasize elements in the same way as the oblique?

Perspective in Pop Culture

As it turns out, the use of the oblique perspective tends to be a technique common to **popular cartography**, that is, the maps of material popular culture.

For example, oblique perspectives of cities and towns, the so-called urban views or bird's-eye views, were mass-produced in the United States during the late nineteenth century. Several thousand American cities and towns were represented in view form by a number of artists, lithographers, and entrepreneurs, many of whom remain anonymous. The views were ubiquitous, displayed on the walls of homes and businesses across the country.

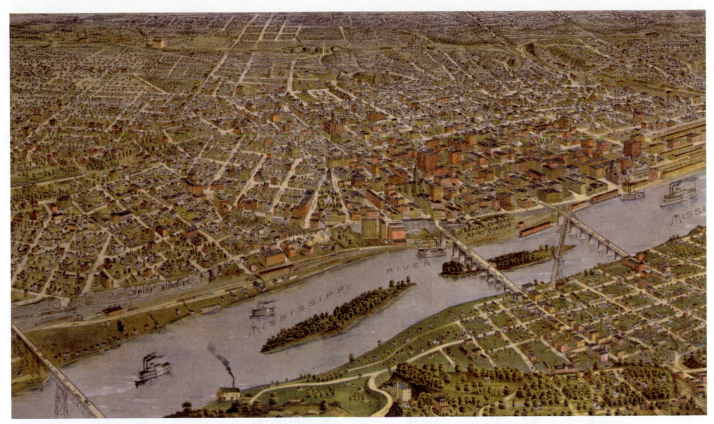

Anonymous. *St. Paul, Minn., January, 1888.* J. H. Mahler Co., 1888. Minnesota Historical Society (MR2.9–SP1e–p10). Used with permission of Minnesota Historical Society.

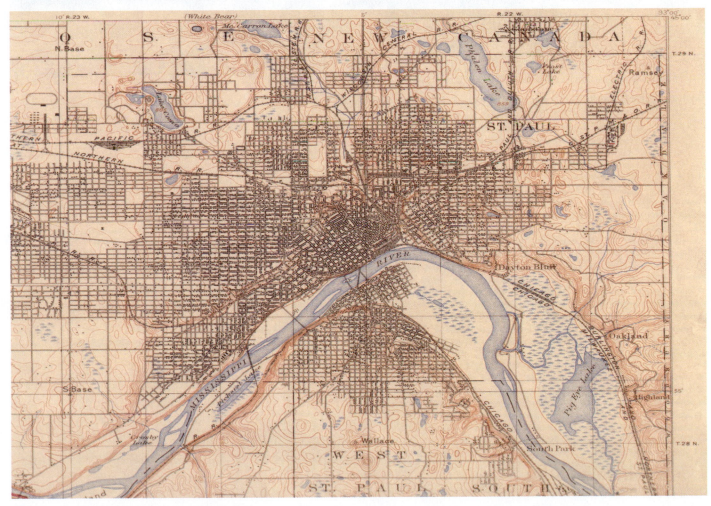

U.S. Geological Survey. *St. Paul.* 1896 ed., rep. 1917.

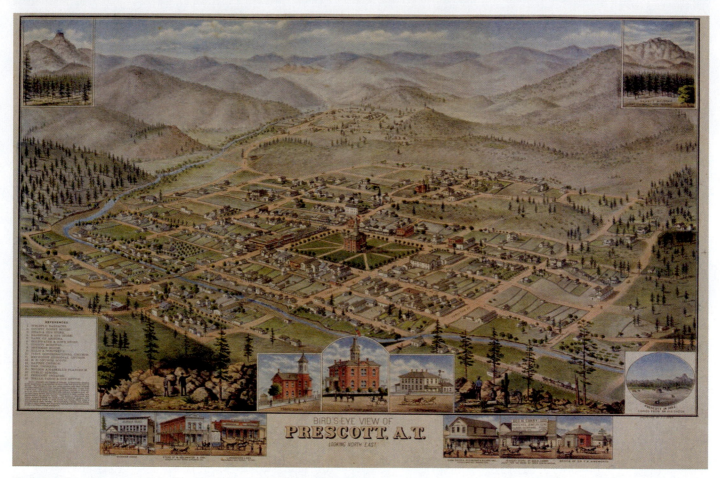

Bird's-Eye View of Prescott, A.T. Looking North East. Chromolithograph, after C.J. Dyer, 1885. Collection of the Amon Carter Museum, Fort Worth, Texas, #1968.45. Used with permission of the Amon Carter Museum.

Above are bird's-eye views for two Western cities, Prescott, Arizona, and Reno, Nevada. Both towns are laid out in a grid structure, with a river running through the town. Their depiction in each of the views, however, is slightly different.

These two views include a third type of perspective, called the **profile view.** A map or image that presents the landscape as if we are standing before it, viewing it at eye level, is a profile view. Although we don't usually think of a profile perspective as a "map," the profile is commonly used in popular cartography in conjunction with the oblique and planar perspectives.

In these two views, the center of the image presents the city in oblique perspective; in the margins, details about the city are presented in profile perspective.

Oblique and profile perspectives seem to depart from the conventional idea of the map. But a careful reading of these two views will begin to reveal the ways in which bird's-eye views could be useful to human geographers.

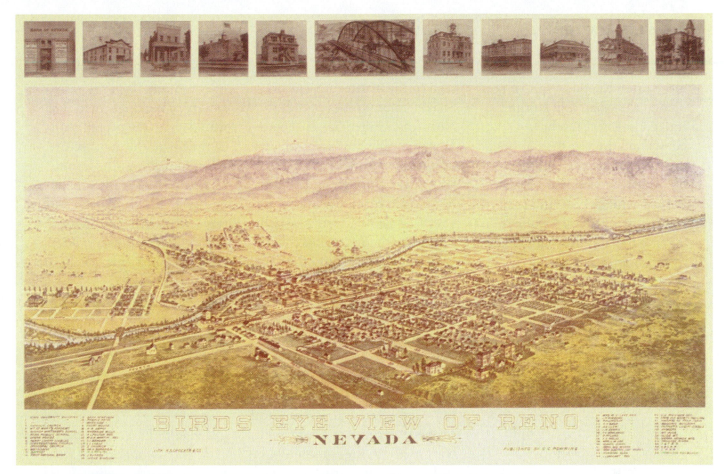

Anonymous. *Birds Eye View of Reno Nevada.* H. S. Crocker & Co.: 1885. Courtesy of Historic Urban Plans, Ithaca, N.Y.

Question 4: What elements of each city are placed in the foreground in these two maps? What elements are placed in the background?

Question 5: What is the difference in the type of information shown in profile in the two views? How does that influence your perception of the city?

● 2.3
Folk Cartography

In Western society, maps are more often the purveyors of popular culture than folk culture. Indeed, many critics have bemoaned the fact that, although the map evokes the landscape of folklife, the folk landscape cannot actually be found there. But one way to find folklife in maps is to go to the maps created by folk cultures.

An interesting example can be seen in the nineteenth-century Shaker communities of the Northeast United States. The Shaker movement began in upstate New York in the 1780s and spread throughout the Eastern United States during the 1800s. Shaker faith emphasized communal living, simplicity, confession, pacifism, and the equality of human beings as the virtues that would bring them closer to God. In their arts, Shakers discouraged decoration for decoration's sake as an unspiritual practice that clouded the spiritual path. Instead, Shaker artisans nurtured a practical, symmetrical aesthetic in order to create works both useful to the community and religiously instructive.

In his book *Shaker Village Views*, Robert P. Emlen provides a comprehensive look at the maps that came out of Shaker folklife. Emlen explains that to manage the communal resources of a Shaker village, a vernacular cartography evolved over time that met the need for both an archived record for the community and a status report to be sent back to the parent ministry.

In each village, one of the men, or "brethren," had the responsibility of making a map that would show the overall plan of the town, the architecture of buildings, and the layout and pattern of planted fields and orchards. To remain within the aesthetic framework of their faith, these mapmakers strove to create detailed, but not decorative, cartography.

On page 23 is an example of a map by Brethren Henry Clay Blinn, known as "Br. Henry," of Canterbury, New Hampshire.

Question 1: Think about the content of the map. What can be gleaned from this document about daily life in the Shaker world? What kinds of agricultural activities could be found in this village?

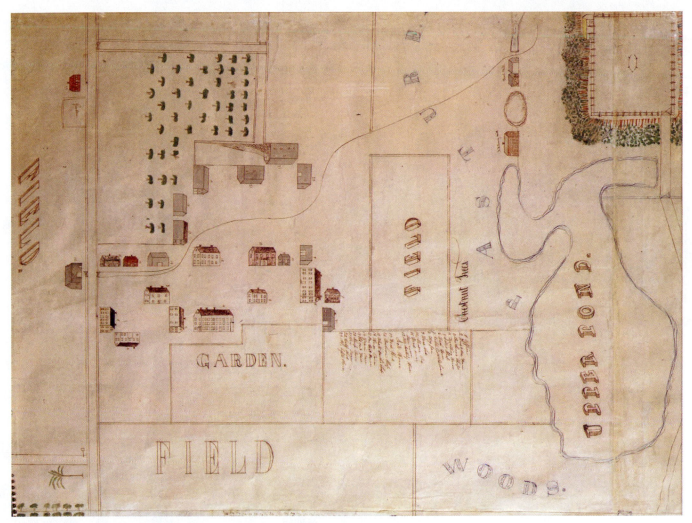

Map of Canterbury, New Hampshire by Brethren Henry Clay Blinn. Courtesy of Canterbury Shaker Village Archives, Canterbury, N.H.
Copywork by Bill Finney, 2001-2002.

Now turn your attention to the way this map was drawn. As in popular cartography, map perspective is often used in folk cartography to communicate certain aspects of geography.

Question 2: How has Br. Henry used map perspective? Is this a planar, oblique, or profile view? How does this differ from what you would expect to see on a conventional topographical map?

Question 3: Why do you think he uses perspective in this way? What elements of the folk landscape are represented as a result?

Question 4: In which direction is the map oriented? Why would this be useful?

Question 5: Shaker artists valued words and images equally as tools for artistic representation. Does Br. Henry use words in his maps differently than the way words are used in an Ordnance Survey map, or a U.S. Geological Survey map? Explain your reasoning.

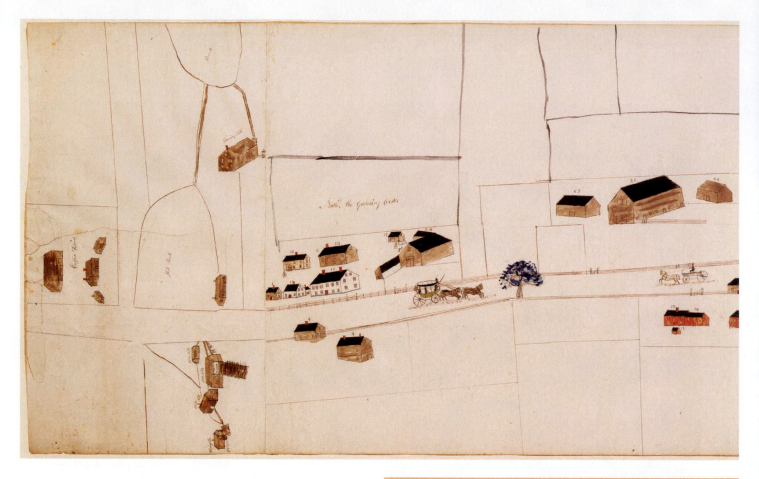

Detail of *Poland, Maine* by Brethren Joshua H. Bussell. Used with permission, United Society of Shakers, Sabbathday Lake, Maine, and Robert P. Emlen.

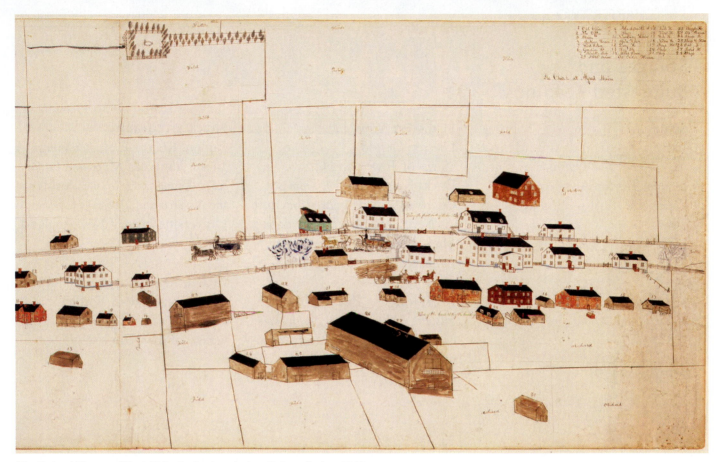

The Shaker Village at Alfred, Maine by Brethren Joshua H. Bussell, circa 1848.
© 2002 Museum of Fine Arts, Boston. Gift of Dr. J. J. G. McCue, 1978.461.

Compare Br. Henry's map to two maps by Brethren Joshua H. Bussell, one depicting the Shaker community at Alfred, Maine, and the other depicting the Shaker community at Poland Hill, Maine.

Question 6: How does Br. Joshua use perspective? How does his use of perspective shape what is depicted for Alfred, Maine?

Question 7: How does Br. Joshua use symbols for geographical features in his maps?

Sources and Suggested Readings

Scale

Larsen, Knud, and Amund Sinding-Larsen. *The Lhasa Atlas: Traditional Tibetan Architecture and Townscape*. Boston: Shambhala, 2001.

Bird's-Eye Views

Danzer, Gerald. "Bird's-Eye Views of Towns and Cities," in David Buisseret (ed.). *From Sea Charts to Satellite Images: Interpreting North American History through Maps*. Chicago: University of Chicago Press, 1990, pp. 143–63.

Reps, John W. *Bird's Eye Views: Historic Lithographs of North American Cities*. N.Y.: Princeton Architectural Press, 1998.

Folk Cartography

Emlen, Robert P. *Shaker Village Views: Illustrated Maps and Landscape Drawings by Shaker Artists of the Nineteenth Century*. Hanover: University Press of New England, 1987.

Ryden, Kent. *Mapping the Invisible Landscape: Folklore, Writing, and the Sense of Place*. Iowa City: University of Iowa Press, 1993.

3

Religion

Vocabulary applied in this chapter
Islam
Christianity
adherents
religious culture region
religious diffusion
pilgrimage
sacred space
Hinduism

New vocabulary
general map
thematic map
dot distribution map
choropleth map
proportional symbol map
aggregated data
enumeration unit
count vs. proportional data
spiritual maps

✹ 3.1
Introducing the Thematic Map

Geographers tend to think of maps as belonging to one of two categories: **general** or **thematic.** A **general map** depicts the locations of a range of geographical features without emphasizing any particular feature. General maps are crucial sources for exploring the cultural geography of a place (the Ordnance Survey maps you studied in chapter 2 are good examples of general maps).

A **thematic map,** on the other hand, emphasizes a theme by showing the distribution of one or two particular features in a region. The thematic map is a fundamental tool for presenting ideas and findings about specific

CNTRY_NAME	%_MUSLIM	CNTRY_NAME	%_MUSLIM	CNTRY_NAME	%_MUSLIM	CNTRY_NAME	%_MUSLIM
Afghanistan	98.1	El Salvador	0.0	Liberia	16.0	Sao Tome and Principe	0.0
Albania	38.8	Equatorial Guinea	4.1	Libya	96.1	Saudi Arabia	93.7
Algeria	96.7	Eritrea	44.7	Liechtenstein	2.7	Senegal	87.6
American Samoa	0.0	Estonia	0.3	Lithuania	0.2	Serbia	16.2
Andorra	0.6	Ethiopia	30.4	Luxembourg	1.0	Seychelles	0.2
Angola	0.0	Falkland Islands (Islas Malvinas)	0.0	Macau	0.0	Sierra Leone	45.9
Anguilla	0.6	Faroe Islands	0.0	Macedonia	28.3	Singapore	18.4
Antarctica	2.4	Federated States of Micronesia	0.0	Madagascar	2.0	Slovakia	0.0
Antigua and Barbuda	0.4	Fiji	6.9	Malawi	14.8	Slovenia	0.1
Argentina	2.0	Finland	0.2	Malaysia	47.7	Solomon Islands	0.0
Armenia	2.7	France	7.1	Maldives	99.2	Somalia	98.3
Aruba	0.3	French Guiana	0.9	Mali	81.9	South Africa	2.4
Australia	1.2	French Polynesia	0.0	Malta	0.5	South Georgia and the South Sandwich Is	0.0
Austria	2.2	French Southern & Antarctic Lands	0.0	Man, Isle of	0.0	South Korea	0.2
Azerbaijan	83.7	Gabon	4.6	Marshall Islands	0.0	Spain	0.5
Bahamas, The	0.0	Gambia, The	86.9	Martinique	0.2	Spratly Islands	0.0
Bahrain	6.3	Gaza Strip	12.0	Mauritania	99.1	Sri Lanka	9.0
Baker Island	0.0	Georgia	19.3	Mauritius	16.9	St. Helena	0.0
Bangladesh	85.8	Germany	4.4	Mayotte	96.5	St. Kitts and Nevis	0.3
Barbados	0.8	Ghana	19.7	Mexico	0.3	St. Lucia	0.5
Belgium	3.6	Gibraltar	8.5	Midway Islands	0.0	St. Pierre and Miquelon	0.0
Belize	0.6	Glorioso Islands	0.0	Moldova	5.5	St. Vincent and the Grenadines	1.5
Benin	20.0	Greece	3.3	Monaco	0.4	Sudan	70.3
Bermuda	0.0	Greenland	0.0	Mongolia	4.8	Suriname	13.9
Bhutan	1.0	Grenada	0.3	Montenegro	16.2	Svalbard	0.0
Bolivia	0.0	Guadeloupe	0.4	Montserrat	0.0	Swaziland	0.7
Bosnia and Herzegovina	60.0	Guam	0.0	Morocco	98.3	Sweden	2.3
Botswana	0.2	Guatemala	0.0	Mozambique	10.5	Switzerland	2.7
Bouvet Island	0.0	Guernsey	0.0	Myanmar (Burma)	2.4	Syria	89.3
Brazil	0.1	Guinea	67.3	Namibia	0.0	Taiwan	0.4
British Indian Ocean Territory	0.0	Guinea-Bissau	39.9	Nauru	0.0	Tajikistan	83.6
British Virgin Islands	0.3	Guyana	8.1	Nepal	3.9	Tanzania, United Republic of	31.8
Brunei	64.4	Haiti	0.0	Netherlands	3.8	Thailand	6.8
Bulgaria	11.9	Heard Island & McDonald Islands	0.0	Netherlands Antilles	0.2	Togo	18.9
Burkina Faso	48.6	Honduras	0.1	New Caledonia	2.7	Tokelau	0.0
Burundi	1.4	Howland Island	0.0	New Zealand	0.2	Tonga	0.0
Byelarus	0.3	Hungary	0.6	Nicaragua	0.0	Trinidad and Tobago	6.8
Cambodia	2.3	Iceland	0.0	Niger	90.7	Tunisia	98.9
Cameroon	21.2	India	12.1	Nigeria	43.9	Turkey	97.2
Canada	1.0	Indonesia	54.7	Niue	0.2	Turkmenistan	87.2
Cape Verde	2.8	Iran	95.6	Norfolk Island	0.0	Turks and Caicos Islands	0.0
Cayman Islands	0.2	Iraq	96.0	North Korea	0.0	Tuvalu	0.0
Central African Republic	15.6	Ireland	0.2	Northern Mariana Islands	0.0	Uganda	5.2
Chad	59.1	Israel	12.0	Norway	1.0	Ukraine	1.7
Chile	0.0	Italy	1.2	Oman	87.4	United Arab Emirates	75.6
China	1.5	Ivory Coast	30.1	Pacific Islands (Palau)	0.0	United Kingdom	2.0
Christmas Island	16.1	Jamaica	0.1	Pakistan	96.1	United States	1.5
Cocos (Keeling) Islands	66.7	Jan Mayen	0.0	Panama	4.4	Uruguay	0.0
Colombia	0.1	Japan	0.1	Papua New Guinea	0.0	Uzbekistan	76.2
Comoros	98.0	Jarvis Island	0.0	Paracel Islands	0.0	Vanuatu	0.0
Congo	1.3	Jersey	0.0	Paraguay	0.0	Venezuela	0.3
Cook Islands	0.0	Johnston Atoll	0.0	Peru	0.0	Vietnam	0.7
Costa Rica	0.0	Jordan	93.5	Philippines	6.2	Virgin Islands	0.1
Croatia	2.3	Juan De Nova Island	0.0	Pitcairn Islands	0.0	Wake Island	0.0
Cuba	0.1	Kazakhstan	42.7	Poland	0.0	Wallis and Futuna	0.0
Cyprus	1.0	Kenya	7.3	Portugal	0.2	West Bank	73.5
Czech Republic	0.0	Kiribati	0.0	Puerto Rico	0.0	Western Sahara	98.0
Denmark	1.3	Kuwait	83.0	Qatar	82.7	Western Samoa	0.0
Djibouti	94.1	Kyrgyzstan	60.8	Reunion	0.0	Yemen	98.9
Dominica	0.2	Laos	0.4	Romania	1.3	Zaire	1.1
Dominican Republic	0.0	Latvia	0.4	Russia	7.6	Zambia	1.1
Ecuador	0.0	Lebanon	42.4	Rwanda	7.9	Zimbabwe	0.7
Egypt	84.4	Lesotho	0.1	San Marino	0.0		

Source: *World Christian Encyclopedia, 2nd ed., Vol. 1: The World by Countries: Religionists, Churches, Ministries.* N.Y.: Oxford University Press, 2001.

topics in cultural geography to an audience. In this section, we will explore the different types of thematic map, and how each affects the way in which we perceive the geographical distribution of culture.

Suppose that the theme you want to map is the global geography of **Islam**. The data set on page 28 shows the percentage of population that is Muslim, both Sunni and Shiite, for each country. In this chart, "0.0" indicates countries that have less than 0.1% Muslim populations, as well as countries for which there was no data for this particular data set.

Question 1: Study the data in the table on page 28. Can you determine from this format which region of the world has the lowest percentage of Muslim population? Why or why not?

Before thematic maps were invented, mapmakers began with a base map of the world, and wrote the number of features inside each of the countries. In our case, the map would look like the one below.

As you can see, the more data you have, the less useful is the technique of placing numbers on maps. We can compare the countries by reading the numbers asso-

ciated with them, but it is difficult to get an overall sense of the distribution of the data set. To get to the visual pattern of the information, geographers eventually replaced the written number with a graphic symbol, creating the tradition of the thematic map.

Despite the seemingly wide diversity of mapping in society, our choices for the thematic map type are somewhat limited. As Arthur Robinson has written, Western mapmakers have discovered only a limited number of ways to show thematic data on maps (all, incidentally, discovered by the end of the nineteenth century).

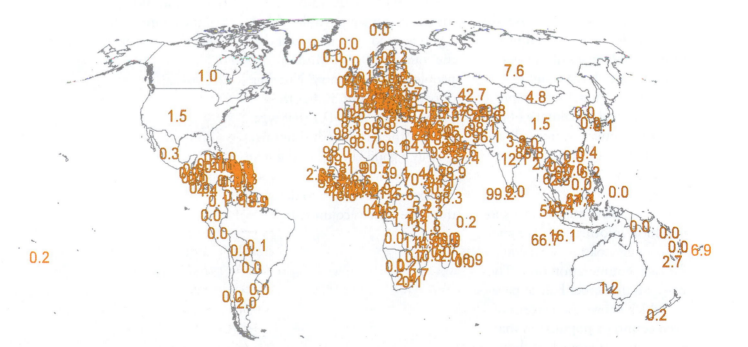

Base map data source: ESRI 1999.

This exercise explores three of these thematic map types: the **choropleth,** the **dot distribution,** and the **proportional symbol.** (Another type of thematic map, the **cartogram,** is explored in chapter 4, and the **isarithmic** map is explored in chapter 7.)

The choropleth, the dot distribution, and the proportional symbol are similar in that they are all useful for displaying data that has been aggregated by enumeration unit. **Aggregated data** means that the data was collected from several locations within a region and then totaled or "aggregated" as a single data value for that region. The **enumeration unit** refers to the geographical unit for which each aggregation in the data set has been collected.

For example, in this exercise, the enumeration unit is a political unit: the countries of the world. Each country has a number associated with it, which represents the total percentage of the Muslim population from all villages, towns, and cities within that country.

The **choropleth map** shows geographical information over an area, using a pattern or color for each enumeration unit. The map on the upper half of page 31, which shows the percent of each country's population that is Muslim, is mapped by choropleth.

Question 2: How does the choropleth improve on the map using only numbers to show the distribution of Muslims?

Question 3: What information is lost?

The choropleth is one of the most popular techniques because it is simple to make with or without a computer. It is, thus, the type of thematic map you are most likely to see in human geography. Ease of construction does not always signify the best solution for mapping a particular geography, however.

For example, a correct choropleth shows **proportional data,** such as percent of population—or per capita—or per square mile. What if you wanted your map to show the total number of Muslim people in each country? Number of people is **count,** rather than proportional, data. For that type of data, the choropleth is not necessarily a good choice. In the map on the lower half of page 31, the choropleth is redrawn to depict the same data as count data.

Question 4: How do the two choropleth maps compare?

Question 5: How does the size of the country influence the data set in the second choropleth?

Question 6: How does the population of the country influence the data set in the second choropleth?

Throughout this workbook, we will be using the choropleth map more than any other map. This is not because the choropleth is the best thematic map for human geography. Rather, it reflects the state of the technology because the choropleth is easier to automate than other thematic map types. With rare exceptions, Web map applications overwhelmingly use the choropleth map over all other thematic map types.

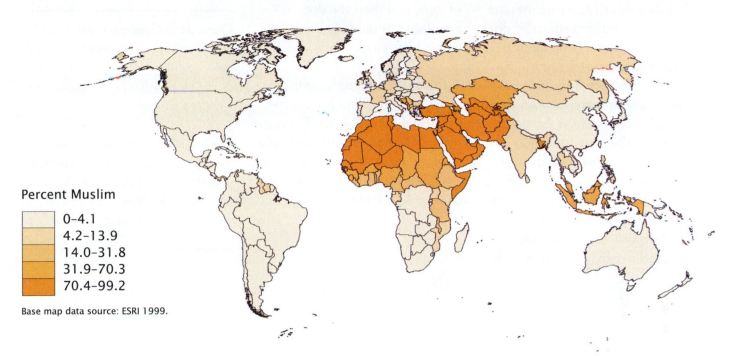

Percent Muslim

- 0–4.1
- 4.2–13.9
- 14.0–31.8
- 31.9–70.3
- 70.4–99.2

Base map data source: ESRI 1999.

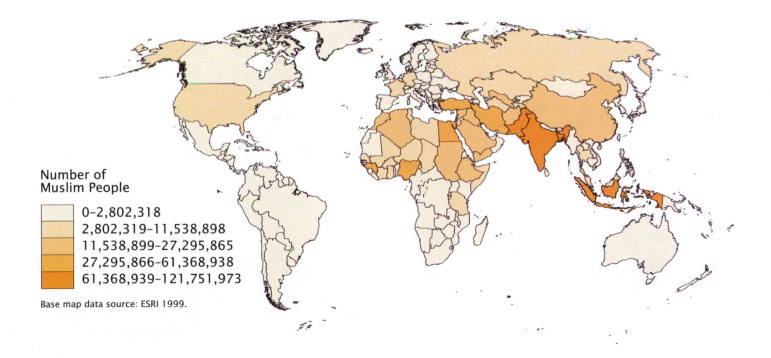

Number of
Muslim People

- 0–2,802,318
- 2,802,319–11,538,898
- 11,538,899–27,295,865
- 27,295,866–61,368,938
- 61,368,939–121,751,973

Base map data source: ESRI 1999.

Another way to show count data is with a **dot distribution** or **dot density** map. As its name indicates, the dot distribution map uses dots to show the distribution of phenomena over an area. The placement of the dots does not show the actual location of the data, however. Dot distribution maps use dots placed randomly across an area to show the relative density of a feature. In the map below, each country is assigned a corresponding number of dots, and then the dots are placed in random locations within that particular country.

A third thematic technique for showing count data is the **proportional symbol map.** This type of map shows the number of phenomena in a particular area using a symbol scaled to represent that data, as in the map on page 33.

Question 7: Compare the dot distribution map, the proportional symbol map, and the first choropleth. How do the maps differ in the spatial distribution pattern depicted in each?

Question 8: Which map do you think is the most useful for a detailed understanding of the geography of Muslim people? On what criteria do you base your opinion?

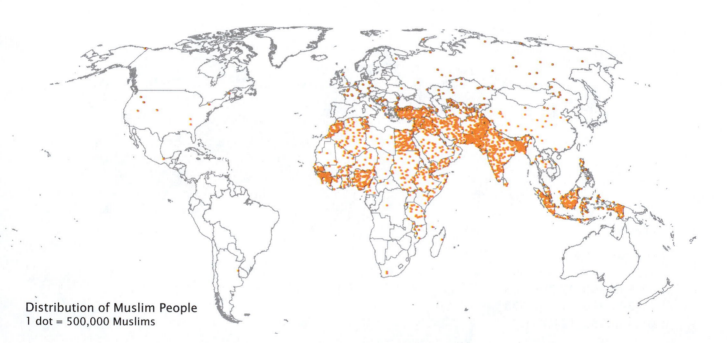

Distribution of Muslim People
1 dot = 500,000 Muslims

Base map data source: ESRI 1999.

Question 9: Which would be the best choice for showing information quickly to an audience in a slide show? Why?

As you can see from the exercise, choosing a good areal thematic map is an art. It takes skill to find the best way to show the information without distorting it and misleading the reader. Like good geographical writing, good thematic mapping takes time and practice.

Number of Muslim People

● 146

11,538,898

103,564,166

Base map data source: ESRI 1999.

⊕ 3.2
Adherents, Membership, and Sects

A map of the geography of religion may be based on any number of indicators of that religion. A map of **adherents** shows the proportion of the population that adheres to the religion, whether or not they have formal ties to a religious institution. A map of membership, on the other hand, depicts those adherents that do have a formal affiliation with a religious institution.

The spatiality of religion can also be understood through mapping locations of objects in the visible landscape: the number and distribution of churches, for example, or the locations of shrines and other holy sites. Geographers try to find the combination of indicators that will best help to uncover spatiality.

Religion is particularly challenging, however, as Edwin Scott Gaustad and Philip L. Barlow note in their *New Historical Atlas of Religion in America*. Gaustad and Barlow note that there is no standard definition of church or other place of worship, nor of adherent, nor member. What makes a church? Is a child an adherent? The more information the map reader has about a particular faith and its records, the more likely an interpretation of the map can be made.

Screen shots © 2002 American Religious Data Archive Web site. Used with permission.

This exercise explores the geography of **Christianity** using an interactive map at the American Religious Data Archive (ARDA). Although the ARDA shows only the geography of Christianity in the United States, the site allows you to explore the spatiality of some 130 Christian sects.

Step 1   Launch your browser and navigate to the American Religious Data Archive at: **www.thearda.com.**

This will take you to the ARDA gateway page, as shown in the screen shot above.

Step 2 In the "Popular Features" box, click on "Interactive Maps and Reports." This will take you to the section on "Church and Church Membership Maps and Reports." You should see a page similar to the middle screen shot on page 34.

Step 3 Under "Mapping," select the "U.S. map" radio button, and click "Go!" You will see a table of variables that can be mapped as in the screen shot at the bottom of page 34.

Take a moment to browse through the religion and other demographic variables in the list. You will notice that for most of the sects, there are multiple data sets available, depending on whether you want to map churches, church members, adherents, or rate of adherence for that sect. Each of these variables alone gives one perspective on the spatial dimension of that religion.

Step 4 Choose a religion from the list of variables that you have learned as having a strong regional basis in the United States. (For example, you may have learned that the Baptist denomination is strongest in the Southeast, Mormons strongest in the West, Catholics strongest in the Southwest, and Lutherans strongest in the upper Midwest.)

Step 5 For the religious variable of your choice, use the "Churches" data set, and click "Go!" You should see a map of the United States depicting this religion by state.

Question 1: Does the map depict a strong religious culture region for the religion you chose?

Question 2: Does the region in the map conform to generalizations about that religion's culture region? How is it the same or different?

Exploring Evidence of Religious Diffusion
Religious cultural regions based on number of churches is a simple geographical concept to visualize, because it is static information with countable geographical features (number of churches). **Religious diffusion,** by contrast, is more difficult to display in a map, because it involves movement, or the interpretation of movement.

The ARDA does not feature animated maps of its religious statistics, but we can get some understanding of diffusion another way: by comparing the number of adherents in a religion to the rate of adherence for that religion.

Step 6 Click the "Back" button on your browser to return to the list of variables. For this exercise we will see if we can detect changes or diffusion in the geography of Friends (the Quakers). Scroll down the list of variables, select "Friends—Adherents," and click "Go!"

Question 3: Describe the regional differences for Friends adherents that you observe in the map. Which region stands out as having the highest number of adherents? The lowest?

Question 4: Does the geography of Friends conform to the broad regional subdivisions of Northeast, Southeast, Midwest, Plains, Southwest, and Northwest? Why or why not?

Step 7 Print a copy of this map using the Print button on your browser. (Even though this map depicts data with colors, if you do not have a color printer you will be able to see the categories clearly in grays.)

Step 8 Return to the list of variables again, select "Friends—Adh Rate" for Friends adherence rate, and click "Go!"

Question 5: Describe the pattern that you see. Where is the rate of adherence high? Where is it low?

Question 6: Compare the map of adherents to the map of rate of adherence, and analyze the difference. Are the maps similar?

Question 7: For the states in the region having high numbers of adherents, are they all increasing in adherents at the same rate? Is there any region that is both low in numbers of adherence and low in adherence rate? How can you account for this?

Question 8: Based on your answers to Questions 5–7, is it possible to generalize about the way in which the Quaker religion is diffusing in the United States? Why or why not?

🌐 3.3
Maps for Pilgrims

Maps are useful for showing the material landscape of the sacred: the locations of shrines and the paths of religious **pilgrimage**.

But maps can also depict the experience of the sacred. Christians, Hindus, Jains, Buddhists, and indigenous peoples around the world all use maps to depict the spaces of spiritual experience.

Spiritual maps are markedly different from both reference and thematic maps. Because the realm of the sacred is often perceived to be incompatible with the realm of the profane, maps of **sacred space** are sometimes composed according to different rules and aesthetic principles than conventional cartographic depictions of the same region. Sacred space, being fundamentally different from profane space, requires a separate cartographic language in order to express its separateness from profane space.

One example of the difference between the religious cartography of pilgrimage and nonreligious cartography comes from the city of Varanasi on the Ganges River in Uttar Pradesh, India.

Varanasi is the ancient "City of Light," the destination for Hindu pilgrims seeking renewal by washing away sins in the sacred Ganges River. Varanasi is also the place where Hindus can be released from samsara, the cycle of life and reincarnation, by dying in the holy space of the city, inside the gates. Known also as the city of Kashi, the name "Varanasi" derives from the names of the tributaries flowing into the Ganges at this place, the Varana and Asi rivers.

In 1876, Kailasanatha Sukula created a map of Varanasi intended to make the city, in his words, "constantly visible for foreigners." Today, a reproduction of Sukula's "Mirror of KÁĐÍ" is on the Web at the Varanasi Research Project—Visualized Space site.

Step 1   Launch your browser and navigate to the Varanasi Research Project at **http://benares.uni-hd.de/**.

You should see the gateway to the Varanasi Research Project Web site as shown below.

Screen shots above and right © 2002 Dr. Jörg Gengnagel, Varanasi Research Project — Visualized Space. Used with permission of the author. Reproduction of the *Mirror of Kashi* by Kailasanatha Sukula used with permission of the British Library.

Step 2 From the yellow bar along the bottom of the Web site, click "The Map." This will take you to a clickable version of the Mirror of KÁÐÍ, as shown in the screen shot below.

At this site, the overall view of the map is shown in the small image in the lower right area of the page.

Step 3 Roll your cursor over the small overview map. The cursor will highlight gray sections of the map, each with its own index number. Each of these squares is an area that can be

zoomed in for better detail and displayed in the larger map area in the left side of the Web page.

Step 4 Take a moment to explore Sukula's map by clicking the individual squares and looking at the zooms that result.

In this map of Varanasi, the Ganges River is represented by a wide, textured river flowing across the lower half of the manuscript. Two narrower rivers are depicted flowing into the Ganges; these are the tributaries Varuna (square 62) and Asi (square 42).

The large thin square at the center of the map symbolizes the gates of the city. Inside this square the main temples and holy sites of the city are depicted with a range of map symbols. Surrounding these central features, the city's religious sanctuaries radiate in a circular pattern.

Historian Jan Pieper studied Sukula's map and compared it to a British map created fifty years earlier by the surveyor and artist James Prinsep.

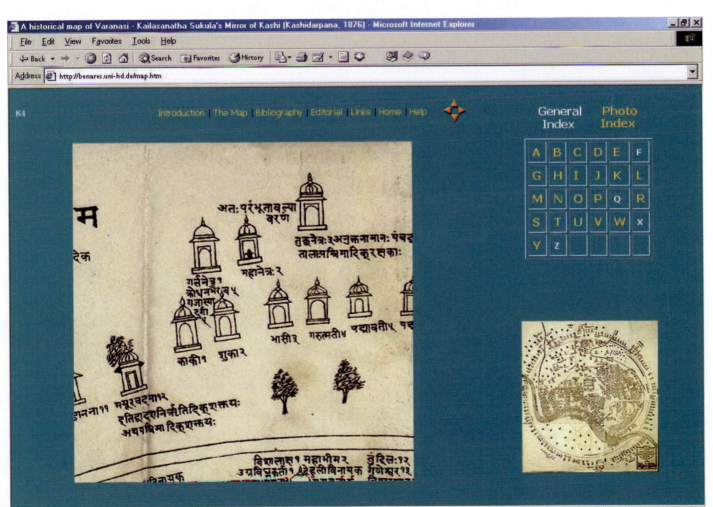

Prinsep's map, shown on page 39, depicts the Ganges River as the eastern border of the city map, with the Varana and Asi rivers flowing into it. Prinsep also depicts the holy sites within the city gates, and lists them by name in the lower right section of the map.

In Pieper's study of the two maps, he points out that although both Sukula and Prinsep intend their maps to be read by visitors to Varanasi, each mapmaker depicts very different spaces of the city, one sacred and one profane, by using very different map symbols or codes.

For example, compare the difference in the way that architecture is depicted in the two maps.

Question 1: Study the symbols used by Sukula. How are buildings depicted? Based on these symbols, which do you think are the most important religious structures in this city? Which do you think are of lesser significance? Why?

Question 2: Compare the way that Prinsep symbolizes buildings. What differences do you observe?

Question 3: How are roads depicted differently in Sukula's map compared with Prinsep's map?

Question 4: How is perspective used in these two maps?

Question 5: What is the difference in the orientation of the two maps? In which direction is Prinsep's map oriented? In which direction is Sukula's map oriented?

The questions above explore the differences between sacred cartography and profane cartography by looking at the contrast in the way features are symbolized. But we can also study the differences by considering which features are included, and which features are absent or not included, in the two types of mapping.

Question 6: Compare the two maps again. What features of the geography of the city are shown in Prinsep's map that do not appear in Sukula's map? Why do you think this is?

"The City of Bunarus" by James Prinsep. British Library Maps 53345(6). Used with permission of the British Library.

Sources and Suggested Readings

The Thematic Map

Robinson, Arthur. *Early Thematic Mapping in the History of Cartography*. Chicago: University of Chicago Press, 1982.

Petchenik, B. B. 1979. "From Place to Space: The Psychological Achievement of Thematic Mapping." *American Cartographer* 6: 5–12.

Adherents, Members, Sects

Gaustad, Edwin Scott, and Philip L. Barlow. *New Historical Atlas of Religion in America*. N.Y.: Oxford University Press, 2001.

Halvorson, Peter L. and William M. Newman. *Atlas of Religious Change in America, 1952–1990*. Atlanta, Ga.: Glenmary Research Center, 1994.

Maps for Pilgrims

Edney, Matthew. *Mapping an Empire: The Geographical Construction of British India, 1765–1843*. Chicago: University of Chicago Press, 1997.

Gole, Susan. *Indian Maps and Plans: From Earliest Times to the Advent of European Surveys*. New Delhi: Manohar, 1989.

Pieper, Jan. "A Pilgrim's Map of Benares: Notes on Codification in Hindu Cartography," *GeoJournal* 3 (1979): 215–18.

Schwartzberg, Joseph E. "Geographical Mapping," in J. B. Harley and David Woodward (eds.). *History of Cartography, Volume 2 Book 1: Cartography in Traditional Islamic and South Asian Societies*. Chicago: University of Chicago Press, 1992, pp. 388–493.

Varanasi Research Project—Visualized Space. Heidelberg: University of Heidelberg—South Asia Institute, 2001. http://benares.uni-hd.de

CHAPTER

4

Language

Vocabulary applied in this chapter
linguistic culture region
core/periphery
language families
linguistic ecology
shatter belt
dialect culture region
toponymy

New vocabulary
bivariate map
quantitative data
qualitative data
cartogram

4.1
Linguistic Regions and Ecology Online

In this section you will explore some of the concepts of linguistic geography using the language data at the *National Atlas of Canada Online*.

First published in 1906 and produced by the Canadian government, the *National Atlas of Canada* is a regularly updated source for visualizing both the physical and human geography of the country.

For the sixth edition, the atlas ceased to be a paper product and is now operated as an online, interactive atlas. One of the first electronic atlases to be freely available over the Web, this atlas also offers more mapping tools and data layers than most online atlases.

Bilingualism and Linguistic Culture Regions

Both English and French are considered the official languages of Canada, with about 90% of the population speaking English or French as their primary language at home, according to the 1996 census.

What does Canada's "linguistic duality" look like? Is bilingualism a separate region, or is it simply the relationship between two formal language regions? This exercise looks at the **linguistic culture regions** of the two official languages in detail.

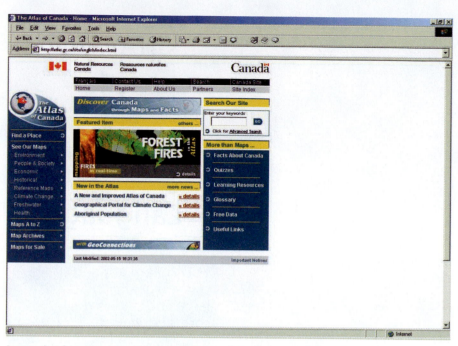

Screen shot from The Atlas of Canada Website at http://atlas.gc.ca, © 2002 Her Majesty the Queen in Right of Canada with permission of Natural Resources Canada.

Step 1   Launch your browser and go to the National Atlas of Canada Online at **http://atlas.gc.ca/english/index.html.**

Step 2   In the left column, under "See Our Maps," choose "People & Society – Official Languages – Knowledge of English." This will launch your first linguistic map of Canada.

Step 3   Study the map that results. The Legend to the right of the map explains the meaning of the colors and symbols used in the map.

(Notice that for this data set, the data is gathered only at populated areas; places without significant populations, and therefore no census data, are shown in grey.)

To get a closer look at the data set, you can zoom in and out using the zoom magnifying tool buttons on the tool bar above the map.

Question 1: Describe the core/periphery pattern of the formal culture region of English in Canada.

Step 4   Return to the list of available layers in the left column, and choose "People & Society – Official Languages – Knowledge of French."

This will redraw your map with "Knowledge of French" as the only visible layer.

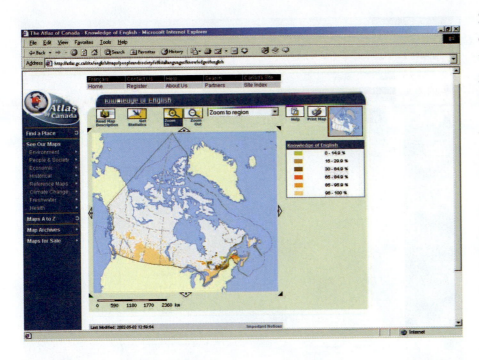

Two screen shots above from The Atlas of Canada Website at http://atlas.gc.ca, © 2002 Her Majesty the Queen in Right of Canada with permission of Natural Resources Canada.

Step 5 Now, repeat the process from Step 4 above to add the layer "English-French Bilingualism," and study the resulting map.

Question 3: How is English-French bilingualism regionally distributed?

Question 4: Is there a strong correlation between the core areas of bilingualism and the peripheral areas of French knowledge or English knowledge?

Question 5: Explore the map more closely, considering rural areas versus urban areas. Is there a correlation between the urban or rural character of a place and the level of bilingualism?

Question 6: Is the identity of a populated place as a coastal or inland place a factor in bilingualism? Is proximity to the United States a factor?

Linguistic Ecology and the Bivariate Map

The Atlas of Canada Online also allows us to look at different data sets simultaneously to see how different linguistic variables are related. This type of data comparison can be used, for example, to explore concepts of **linguistic ecology**. Although linguistic ecology sometimes refers to a wide range of contextual variables influencing language (such as community size and political issues), in this exercise we will explore it as the relationship between language and physical geography.

Because the "Official Languages" data set is symbolized as a choropleth map each time, only one layer of data is visible at a time.

Question 2: Describe the core/periphery pattern of the formal culture region of French. How is it different from the English culture region?

Step 6 Return to the list of layers in the left column, and select "People & Society – Aboriginal Languages – Aboriginal Languages by Community, 1996." The map that you see should look like the image on the right.

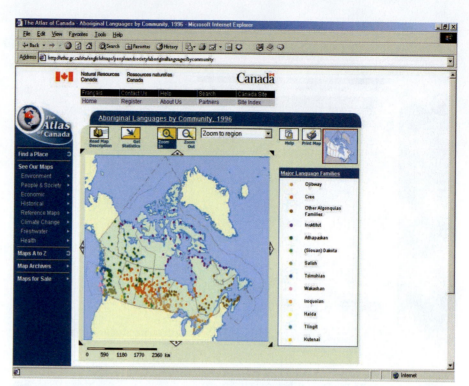

The Atlas of Canada Website http://atlas.gc.ca © 2002 Her Majesty the Queen in Right of Canada with permission of Natural Resources Canada.

Question 7: Study the linguistic differences shown in this small-scale map. What region of Canada has the greatest diversity of aboriginal languages? What region has the least diversity?

Question 8: Review the connection between landscape and linguistic diversity that you have learned in class and reading. What would you expect the landscape to be like in the region with the highest density of language families? the lowest density?

Step 7 Zoom in on any region of the map by clicking the "Zoom in" magnifying tool on the tool bar, and clicking your cursor on the map to activate the zoom. You will see that the point symbols show not only the language spoken, but also the number of speakers at that point.

The depiction of two data variables in one map is called a **bivariate map.** Any type of thematic map can be bivariate.

Question 9: Consider the way in which the two data sets are symbolized in this online map. What type of thematic map is this?

The bivariate symbols in this map are combining two major types of data: **quantitative data,** which is ranked or numerical, and **qualitative data,** which is categorically different and therefore cannot be ranked.

The size of the circle shows the quantitative data, in this case— the population size of the community. The color or hue of the point symbol shows the qualitative data, the major language family for that community. Both types of symbols are explained separately in the legend to the right of the map.

Step 8 Explore the aboriginal language geography by panning around the map. Compare some of the communities by clicking the "Get Statistics" button above the map, and clicking on one of

the circles. This will take you to a pop-up window with data for the number of speakers in the community. If you click on the link for "Aboriginal Community Statistics" in the pop-up, you can delve into actual population numbers as well as socioeconomic data for the community.

As you explore, you will notice that there are strong regional differences in language families, community populations, and number of speakers.

Is there a close connection between linguistic regions, number of speakers, and physical geography? In the next section, we will explore this question by zooming in to the region of southern British Columbia and Alberta.

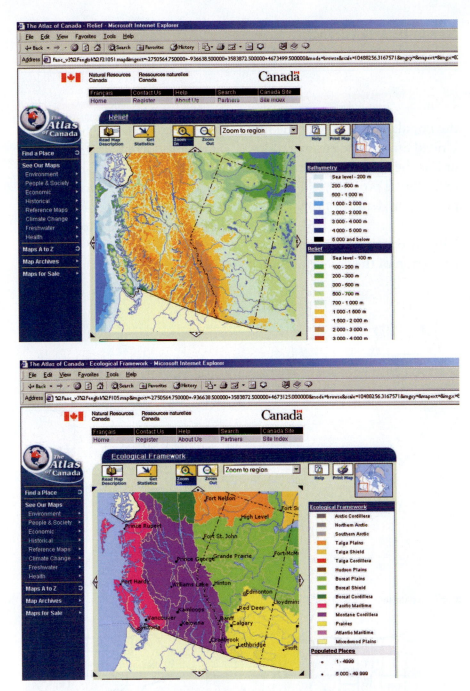

Two screen shots above from The Atlas of Canada Website at http://atlas.gc.ca, © 2002 Her Majesty the Queen in Right of Canada with permission of Natural Resources Canada.

Question 8: What are the majority language family or families in use in this region?

Question 9: Compare the geographic information that you see in the linguistic map with the maps of relief and ecological framework, left. Can you find evidence for a linguistic shatter belt? Explain your findings.

Step 10   Analyze both the population and number of speakers information in this region by comparing the statistics for the larger communities with the statistics for the smaller communitites. Compare your findings to the maps on the left.

Question 10: Is there a relationship between number of speakers and land relief? Between number of speakers and ecological framework?

Question 11: What is the relationship between number of speakers and size of the populated place? Why do you think this is?

Question 12: What seems to be the typical size of the populated place where an aboriginal language family is located?

Although this atlas includes physical mapping layers such as relief and ecological information, these layers cannot be viewed simultaneously. To help you make comparisons, the data layers for Relief and Ecoogical Framework (from the "Environment—Land" section of the atlas) are provided in the two figures, above.

Step 9   Zoom to the region of Southern British Columbia and Alberta. Explore the map using your pan and zoom tools.

✴ ⊕ 4.2
Locating Vernacular Dialect with the Cartogram

What do the **dialect culture regions** of the United States look like? How does map type affect how we see these regions?

These are questions that are addressed in the *Dictionary of American Regional English* (*D.A.R.E.*) project, the purpose of which is to index and map the vernacular dialect of the United States. The dictionary is based primarily on verbal sources, using field interviews conducted in 1002 communities from 1965 to 1970. The dialect data from these interviews were then supplemented with dialect information from written (published literature) sources.

Looking at Cartograms
The *D.A.R.E.* project uses a particular type of thematic map to display dialect vocabulary, which is called a cartogram.

A **cartogram** is a map in which the size of the enumeration units is proportional to the data represented by those units. Cartograms have no map projection; they are "projected" by the data itself.

In the case of the *D.A.R.E.* cartograms, each cartogram shows the population density of the fifty United States. The enumeration unit is the state, so the size of each state is proportional to the population density of that state.

Directly below is an equal-area projection of the conterminous United States, in which states are shown proportional to their physical size. Below that, for comparison, is the *D.A.R.E.* cartogram, with states sized proportionally to their population density.

Base map data source: ESRI 1999.

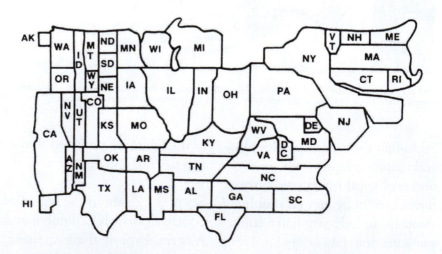

Reprinted from the *Dictionary of American Regional English,* edited by Frederic G. Cassidy, Cambridge, Mass.: The Belknap Press of Harvard University Press. Volume I, A-C, © 1985 by the President and Fellows of Harvard College. Used with permission of the publisher.

Question 1: Compare the regions of the Northeast and West. How do they compare in size in the two different maps?

The *D.A.R.E.* project applied this population density cartogram as the base for maps of the geographical distribution of the vernacular dialect words they recorded.

For example, when asked to name "the small insect that flies at night and flashes a light at its tail," many of the interviewees responded "firefly." The cartogram for "firefly" is shown in the first map, below.

Each black dot on the cartogram represents one interview in which the person used the word "firefly."

If everyone interviewed used the word "firefly," the word would be simply a standard vocabulary word, not part of regional dialect, and the map would look like the second map, below.

Question 2: In which states would you expect the word "firefly" to be relatively common?

Question 3: In which state does the word "firefly" appear to be an anomaly?

Question 4: Based on your interpretation of this map, would you say that "firefly" is regionally differentiated?

Question 5: How does the cartogram influence your interpretation of the usage of this word? What kind of visual picture do you think you would get if the dots were plotted on a conventional map like the one on page 46?

Question 6: Can the map above left be considered a dot distribution map? Why or why not?

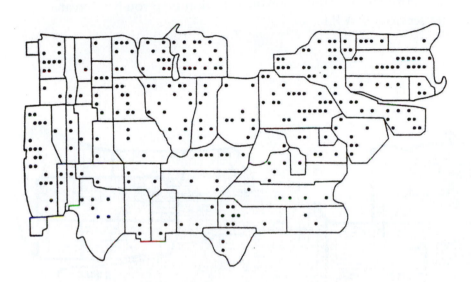

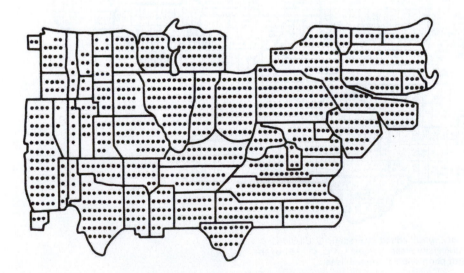

Making Your Own Dialect Maps

Instead of "firefly," eight of the people interviewed gave the word "candle fly." Does candle fly have a regional basis? One way to find out is to make your own vernacular map of this word and look at the distribution.

Each of the "informants," or persons interviewed, was assigned a number to help identify their community inside a state. The map below shows the schematic placement for the numbers.

For this exercise, a blank version of the *D.A.R.E.* map has been created for you to use as a base.

Step 1   Make a photocopy of the blank informants map on page 49.

Step 2   For each of the candle fly informants, place a black dot in the square corresponding to that informant. Use the following data set from *D.A.R.E.*:

Informants who responded "candle fly":

State	Informant Number
AR	41
FL	35
VA	26
LA	14
LA	15
MO	9
NC	10
NC	27

Step 3   Study the dialect distribution map you have created.

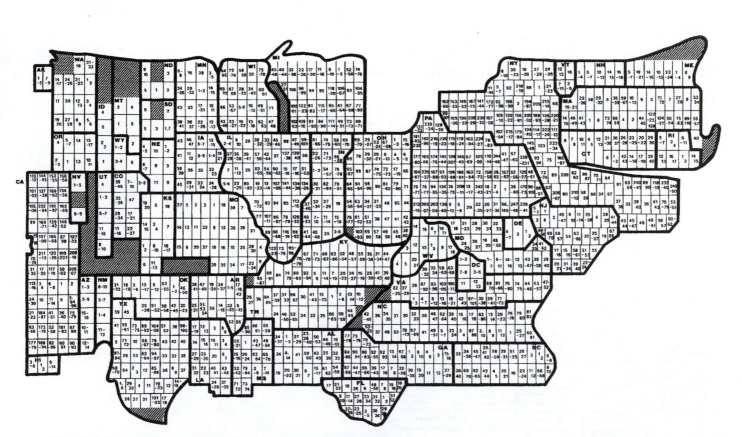

Reprinted from the *Dictionary of American Regional English*, edited by Frederic G. Cassidy, Cambridge, Mass.: The Belknap Press of Harvard University Press. *Volume I, A-C,* © 1985 by the President and Fellows of Harvard College. Used with permission of the publisher.

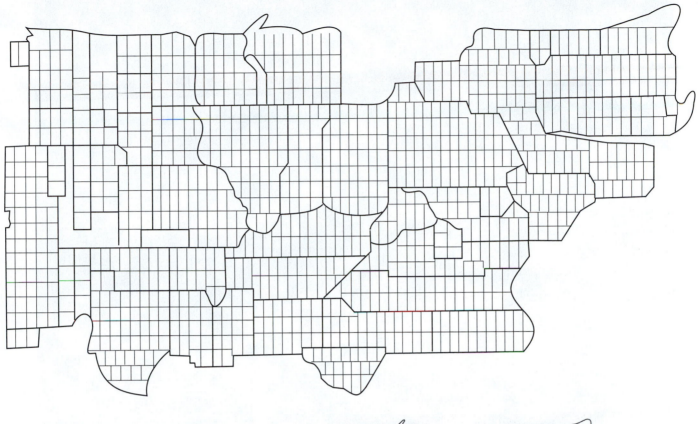

Question 7: *What is the regional distribution of "candle fly"?*

Question 8: *Does the population density of the state seem to be related to the spatial distribution of this word?*

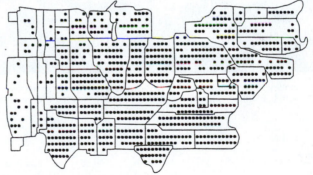

Step 4 Compare your candle fly cartogram to two other variations for the word "firefly": "lightning bug" (above, right) and "firebug" (below, right).

Question 9: *Based on your interpretations of the cartograms for lightning bug, firefly, candle fly, and firebug, explain the regional dialect pattern of "the small insect that flies at night and flashes a light at its tail...."*

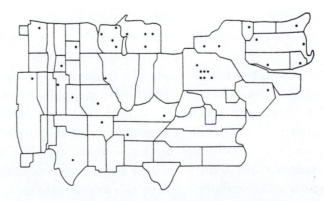

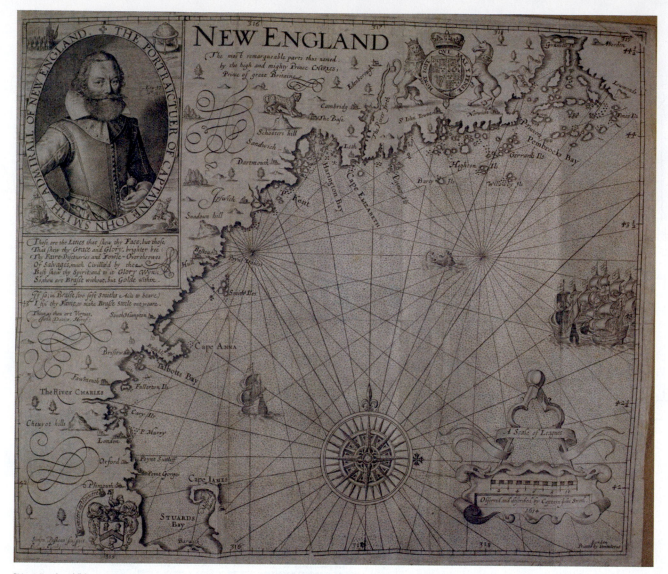

"New England," by John Smith. In *A description of New-England: or The observations and discoueries, of Captain Iohn Smith....* London 1616. Collection of the William L. Clements Library, University of Michigan. Used with permission.

🐾 4.3
The Power of Place Names

Maps can powerfully shape our cultural worldviews. Part of that power stems from the place names associated with particular geographic features, also called **toponymy.** The language of place names, the familiarity or unfamiliarity of their sound and spelling, and even the number and density of named features all contribute to the way in which we perceive the geography of a place.

What do place names on a map tell us about the languages actively spoken in a particular place? Sometimes, very little. Much of cartography stems from the outside looking in, and may as much portray the linguistic hopes or ambitions of the map-maker as the linguistic reality of the place in the map.

In fact, maps are so useful for replacing one culture's toponymy with another, that they have been put to use as traditional tools for the colonization and domination of populations.

For example, consider the 1616 map by John Smith, above. Smith used this map to depict the northeastern region of North America, the theater of his northern explorations, for European audiences. At that time, the region was wholly an indigenous landscape, shaped, cultivated, named, and mapped by native people. The map, however, portrays a land filled with English toponyms, though the first English colonists would not begin arriving at Plymouth for another four years. On his map, Smith notes that the Prince of Wales provided the place names.

Question 1: Why do you think Smith would depict the region in this way, knowing that there was yet no English settlement in the area?

Question 2: What do you think are the long-term effects of removing indigenous place names during colonization?

Today, returning indigenous place names to the map is a priority in places around the world that are undergoing decolonization, or where there is active resistance to colonization.

In Mexico, Guatemala, Wales, Ireland, Canada, and the United States, cultures that have been linguistically colonized are now adding or replacing their own toponyms on maps depicting their territory. By remapping the indigenous words, they are reclaiming their linguistic imprint, and thus their cultural imprint, on the land.

One person who is working to preserve place names through mapping is the Irish cartographer Tim Robinson. Robinson began walking and exploring the Aran Islands on the western coast of Ireland in 1972.

The only available map of the islands at that time was a relic of the British Ordnance Survey mapping of Ireland during the nineteenth century. In that survey, British surveyors set down Irish toponymy with anglicized words that sounded similar to the original Irish words, creating new, colonized, toponymic landscapes on the maps.

Disappointed in the place names provided on his Ordnance Survey map, Robinson set out to remap the original Irish names in the landscape himself, by walking and asking questions of people he met along the way.

He began with the Island of Árainn, and the result, the southeast portion of which is shown on page 52, is a map and text brimming with Irish place names and stories never before set down in print. In his map, numbers refer to stories and additional place names described in a supplementary text.

Robinson's map differs greatly from the early OSI (Ordnance Survey of Ireland) map of this area, shown on page 53. The top map also depicts the southeast section of the Island of Árainn. The two images below it are close-ups from two sections of the map, to give you an idea of the place name geography.

Question 3: Compare the nature and distribution of the place names in the OSI map with those in Robinson's map. In what ways do they differ? Are there any similarities?

Question 4: Assuming that you have not traveled to Árainn, what kind of mental picture of this place do you form from the OSI map? What would you expect the physical landscape to be like? What would you expect of the cultural landscape?

Question 5: Does the Robinson map give you the same picture? Why or why not?

Question 6: In what way is physical landscape connected to the life of a language? Do mapmakers play a role in this connection?

Detail from "Oileáin Árann" by Tim Robinson, © 1996 Folding Landscapes. Used with permission of the author and publisher.

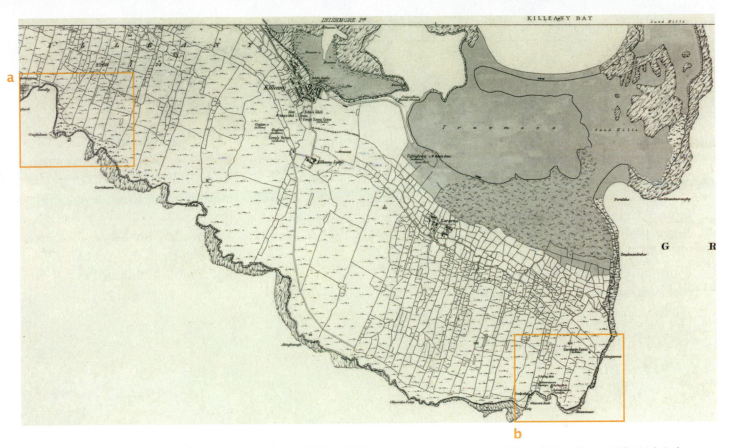

a

b

Details above and left from "County Galway Sheet 119," British Ordnance Survey of Ireland, 1901. Courtesy of the Geography and Map Division, Library of Congress.

Sources and Suggested Readings

Multivariate Explorations

Monmonier, Mark. *Mapping It Out: Expository Cartography for the Humanities and Social Sciences.* Chicago: University of Chicago Press, 1993.

Monmonier, Mark. *How to Lie with Maps.* Chicago: University of Chicago Press, 1991.

Locating Vernacular Dialect

Cassidy, Frederic G., ed. *Dictionary of American Regional English.* 3 vols. Cambridge, Mass.: Harvard University Press, 1985–2002.

Irish Place Names

Robinson, Tim. "Oileáin Árann." [map] 1:28,160. Roundstone, Ireland: Folding Landscapes, 1996.

Robinson, Tim. *Setting Foot on the Shores of Connemara & Other Writings.* Dublin: Lilliput Press, 1996.

CHAPTER
5

Ethnicity

Vocabulary applied in this chapter
U.S. Census
ethnic group
ethnic homeland
ethnic island
ethnic neighborhood

New vocabulary
classification scheme
census tract
equal intervals scheme
class breaks
class width
percentiles scheme
outliers

🔍 5.1
Census Tools for Mapping Ethnic Diversity

This exercise explores the geography of ethnicity across the United States using the online mapping tool of the **U.S. Census,** American FactFinder. In addition to basic population numbers, the U.S. Census collects a variety of socioeconomic data useful to cultural geographers. American FactFinder makes some of this data freely available on the Web.

Step 1   Launch your Web browser, and go to the U.S. Census home page at **www.census.gov.** This address will take you to the main site for all census products available on the Web.

Step 2   Click on "American FactFinder" in the left column.

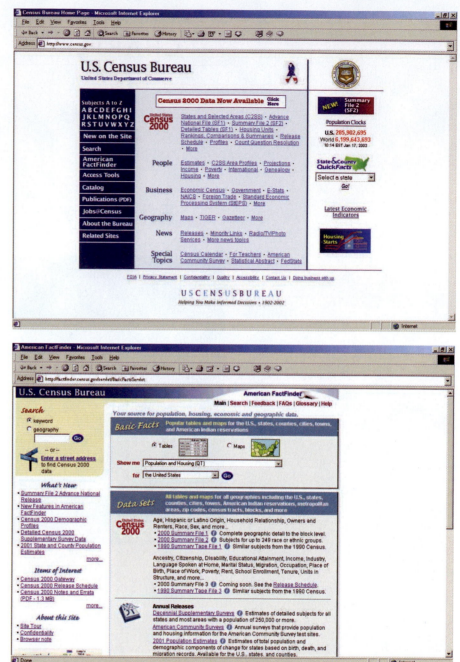

Screen shots from U.S. Census Bureau Web site, American FactFinder (http://factfinder.census.gov), 2001.

FactFinder allows you to explore a myriad of socioeconomic data by tables, maps, or both. Occupation, number of foster children in the family, travel time to work, and language spoken at home, are just some of the variables accessible here. Which particular data is available depends on the geographic scale that you are viewing (for example, state, county, or metropolitan areas) and whether the type of display you would like to view is a table or a map.

For this exercise, we will focus on the mapping function to help visualize spatial differences in **ethnic groups**.

Step 3 Scroll down to the Maps section, and click on the selection "Thematic Maps." You are now at the main mapping page of FactFinder, which displays a map of "Persons per Square Mile: 2000" for the United States.

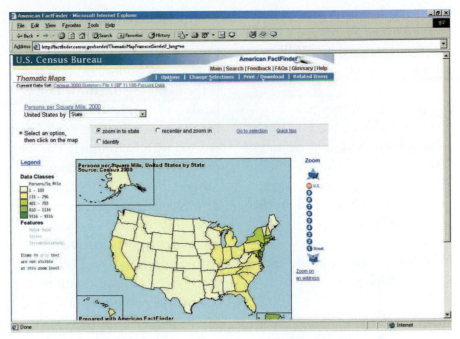

U.S. Census Bureau Web site. Census 2000, Summary File 1. American FactFinder (http://factfinder.census.gov), 2001.

To get into the ethnicity data, you first need to change the data set displayed in the map. Notice that in the upper left corner of the Web page, there is a note that indicates that you are viewing "Current Data Set: Census 2000 Summary File 1 (SF 1) 100-Percent Data." FactFinder allows you to browse a number of different census data sets, one of which is "Summary File 1." Each data set shows different kinds of population information. In Census vocabulary, different types of data within a data set are called "themes" (hence "thematic maps").

Step 4 From the blue menu bar across the right corner of the Web page, roll your cursor across "Change Selections" and select

"Data Set," which will take you to a list of data sets available for mapping. You can get a list of themes within each data set by highlighting the data set in the list, and clicking on the "What's This?" button.

For this exercise, we want to look at the ancestry themes, which are in the data set Summary Tape File 3 (STF 3). As you can see from your data set list, STF 3 is available only on American FactFinder for 1990, from the 1990 Decennial Census of Population and Housing.

Step 5 Choose "1990 Summary Tape File 3" and click the "Next" button.

Next, you need to specify where in the United States you would like to view this data. We'll start with the country as a whole.

Step 6 From the "Select Geography" Web page, leave the selection method at its default of "list." For geographic type, select "Nation." For geographic area, select "United States," then click the "Next" button.

You should now be at the "Select Theme" Web page.

Step 7 Choose "Search...by subject," go to the "Population Totals–Race and Ethnic Groups" subject category, choose Ancestry, and click "Search." The search will return a list of all available ancestry themes to portray in the map.

Step 8 Scroll through the list of themes and choose one that is most interesting to you. When you have selected it from the list, click the "Show Map" button.

You should now be looking at the distribution of the ancestry group by state for the whole United States.

Ethnic Homelands and Ethnic Islands

One aspect of ethnic geography that we can explore from here is the existence of **ethnic homelands** and **ethnic islands**. Examine the generalized map you have created with FactFinder. The map probably shows some regional differentiations, but because the data is aggregated to the state level, it is not a very good source for analyzing the locations of ethnic homelands and islands. For better detail, you want to change the enumeration unit from states to counties.

Step 9 In the upper left portion of the Web page, below the name of the ancestry theme, you will see a menu indicating that the mapped data is "United States by...State." Scroll down the menu and change the selection to redraw the data as "United States by...County."

Step 10 Begin exploring the map for evidence of an ethnic homeland, or ethnic islands for this ancestry group.

Tips for exploring: If the black lines of the county boundaries are interfering, try zooming into level 8 or 9 using the Zoom buttons on the right. You can then pan around the map using the pan arrows at the edges of the map.

If the roads or water features are also interfering, you can turn off these features by going to the left side of the page, above the data classes, and clicking on "Legend." In the Legend box, select "Features," turn off the features you don't want, and click "Update."

To identify a particular county, click on the "identify" radio button from the grey menu above the map, and then click on the county you want to identify.

Step 11 Analyze the spatial pattern that you observe in the map.

Question 1: Which ancestry group did you choose? Does this population appear to have an ethnic homeland? If so, describe its location. If you don't think there is an ethnic homeland evident, what characteristic about the spatial pattern in the map gives you that impression?

Question 2: Where are the ethnic islands for this ancestry group? Describe their size, and whether they appear to be associated with urban or rural regions.

Step 12 When you are finished analyzing the locations of homelands and islands, leave the Factfinder Web site by quitting your Web browser.

Close-up on Florida

In this next exercise, we will focus on ethnic patterns in Florida, using FactFinder to explore differences in spatial distribution.

Note: If your Web browser is running, quit the browser now. This exercise will work best if Factfinder is reset by quitting and restarting the Web browser software.

Step 13   Return to the U.S. Census Web page at **www.census.gov**, and click on FactFinder and "Thematic Maps" as before. From the default population map on the FactFinder page, change the Current Data Set selection again by going into the "Change Selections...Data Set" menu, selecting "1990 Summary Tape File 3," and "Next." (This is a repeat of Steps 1–5, above.)

Step 14   From the "Select Geography" web page, leave the default at "list." For "Select a Geographic Type," choose "State," select "Florida" from the list, and click "Next."

Step 15   For "Select Theme," choose "by subject," "ancestry," and "search." This time, change the ancestry theme to "Percent of Persons of Greek Ancestry," and then "Show Map."

Step 16   Zoom to level 8 so you have a good close-up of the state.

Question 3: Describe the spatial distribution that you see for Greek ancestry. Where are the ethnic islands located?

Step 17   Continue to explore the ancestry distribution by creating and interpreting maps for three additional ancestry categories: West Indian Ancestry, Italian Ancestry, and United States Ancestry. To change the ancestry theme, return to the upper right blue menu bar and choose "Change Selections - Theme," and then "by subject - Search," as you did in Step 15.

Step 18   Analyze the spatial patterns for West Indian, Italian, and United States ancestries.

Tips for analyzing the spatial patterns: When making comparisons between groups, be sure to take into account the different proportions of each ancestry group.

Notice that the themes are classified differently in each map; dark green may represent 2% of the population in one map and 40% of the population in another.

If it helps remind you of the distributions, you can print a copy of each map using the "Print / Download" menu from the blue toolbar in the upper right corner of the page.

Also, if the map is too crowded to read, you can turn map features on or off to assist with your interpretation.

Question 4: How do the spatial patterns of these next three ancestry groups compare to the first? In what ways are they different or the same?

✳ ⊕ **5.2**
Hiding and Finding Ethnic Neighborhoods with the Choropleth

Each of the maps you created in the exercise above is a choropleth map. As you learned in chapter 3, a choropleth is a map that symbolizes data over an area using color or pattern. The choropleth is a popular mapping tool because it is simple to construct digitally and easy for general audiences to read. Because choropleths are ideally suited to mapping data aggregated by enumeration unit (by county or by zip code, for example), it is often the map chosen by geographers to show population characteristics such as income, age, gender, and ethnicity.

A choropleth map assigns one color to an area, as if that area were actually a point. Only one data value can be portrayed in a simple choropleth map because one color is assigned to one value. For example, in the ancestry maps you created with Fact-Finder, you could view only one ancestry theme at a time.

If the data to be mapped is one of many different variables, the result is that the choropleth hides as much data as it shows, which can mislead the reader into thinking that there is only one variable. This is an old problem that has been around as long as the thematic map itself.

For instance, if you want to make a map showing the dominant ethnicity for each **ethnic neighborhood** in your city, you would assign one color per neighborhood depending on which group is dominant. All other ethnic groups are hidden.

Would such a map give an accurate portrayal of ethnic diversity? How would you portray the second most dominant ethnic group? For a map of ethnic diversity, the choropleth hides more information than it displays.

What is the solution?

Cartographer Eugene Turner pondered this problem of the hidden data of the choropleth in 1989, while searching for an accurate portrayal of ethnic diversity in Los Angeles. His solution was to make three separate maps of the city, shown on this page.

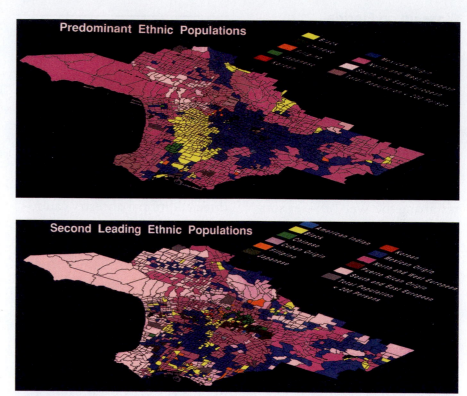

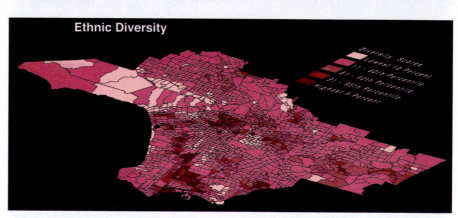

"Leading Ethnicity" (top), "Second Leading Ethnicity" (middle), and "Ethnic Diversity" (bottom),
© 1980 Dr. Eugene Turner, California State University, Northridge. Used with permission of the author.

Question 1: Study the three Los Angeles maps. Why did Turner choose these particular categories for his map series?

In the 1990s, Bruce Macdonald pondered this same problem while working on his book, *Vancouver: A Visual History*. Macdonald wanted to accurately portray the character of ethnic neighborhoods in Vancouver. He experimented with the technique of using the area symbolism of the choropleth map to indicate the largest concentration of an ethnic group for each neighborhood. Then for each choropleth area he provided both the percentage of the dominant group for that neighborhood as well as the city-wide percentage of that group.

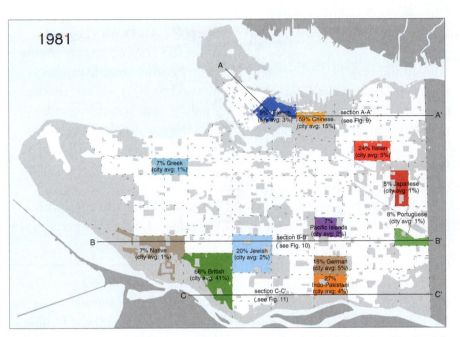

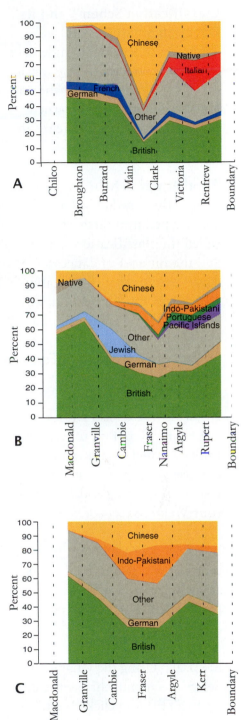

Above: "Largest Ethnic Origin Group by Census Tract, 1981," and right: "Ethnic Cross-Sections" by Bruce Macdonald, from *Vancouver: A Visual History*, Vancouver: Talonbooks © 1992. Used with permission of author and publisher.

This is illustrated in his map of ethnicity in 1981, above.

This technique allowed him to show whether an ethnic group represented a strong majority in a given neighborhood, or merely the greatest proportion in a neighborhood in which no ethnic group held a majority.

To this altered choropleth Macdonald added three transects of race in his city (the paths labeled A, B, and C in the map), to show the changing ethnic neighborhoods as one travels down the individual streets. These transects are shown to the right of the map.

Question 2: Compare the differences in ethnic neighborhoods as shown by the three transects.

Question 3: How do these transects change your perception of Vancouver's ethnic neighborhoods from the perception given by the altered choropleth map?

Question 4: How does Macdonald's portrayal of ethnic diversity compare with Turner's maps? Which do you think is a more effective solution?

Classification Scheme and Its Effect on Small Details

There are other ways to hide data in a choropleth map. One way to do it is to alter the size of the enumeration units, as you learned in the thematic map exercises in chapter 3.

Another way to hide data is to alter the way data values are grouped into categories, called the **classification scheme,** and the number of data categories used in that classification scheme. In this last section, we will explore the effect of different classification schemes on ethnicity data using the tools of Fact-Finder.

Note: As in Exercise 5.1, for best results it is imporant to reset Fact-finder before beginning a new exercise using new data. You can reset it by quitting and restarting your Web browser software.

Step 1   Launch your browser, navigate to the Census at **www.census.gov,** and select FactFinder Thematic Maps as you did in exercise 5.1.

Step 2   Using the Change Selections menu, change the data set to STF 3 and click "Next."

Step 3   Leave the selection method at the default "list." Set the geographic type to County, set the state to California, and set the geographic area to Los Angeles County. Click "Next."

Step 4   From the Select Theme page, search "by subject," select "Ancestry" and click "Search." Choose the Russian ancestry theme and click "Show Map."

You should now be looking at a map of Los Angeles County showing Percent of Population with Russian Ancestry, 1990.

Step 5   Zoom to level 7 and recenter the map on the county.

For this exercise, we will explore the data at the level of census tract for the enumeration unit. In U.S. Census data, a **census tract** is an enumeration unit smaller than the county level, averaging about 4,000 people in size.

Step 6   Change the enumeration unit to census tract using the drop-down menu above the map viewer. Change it to specify "Los Angeles County, California by—Census Tract."

Now that the enumeration unit is set, what should the classification scheme be? One of the most popular schemes is **equal interval** classification because it is easy to construct. In equal interval classification, the all data values in the data set are divided into categories (called "classes") of equal numerical value.

The values which begin and end each of the data categories are called the **class breaks.** The differences from the highest value to the lowest is called the **class width.** In equal interval schemes, each of the data categories has the same class width.

Step 7   Click on "Legend" to set the graphic display of the map. Set the variable ranging method to 10 intervals, choose a palette that you think will effectively symbolize low to high values, and create a GIF image map.

Step 8   Print this map with a legend, and then quit FactFinder.

Question 5: Analyze the map you printed. What, in general, is your perception of the geography of Russian Ancestry in Los Angeles County?

Question 6: What are the changes like between census tracts? Is there a smooth or abrupt change in percent ethnicity?

Question 7: Do you think equal intervals are a useful way to categorize Russian Ancestry? Why or why not?

In their atlas of ethnicity and socioeconomic status in Southern California, *The Ethnic Quilt,* James P. Allen and Eugene Turner mapped the spatial distribution of more than 30 individual ethnic identities. To achieve this, the maps in their atlas show the spatial distribution of a single ethnicity as a choropleth map, at the census tract level, using another type of classification: the **percentiles scheme.**

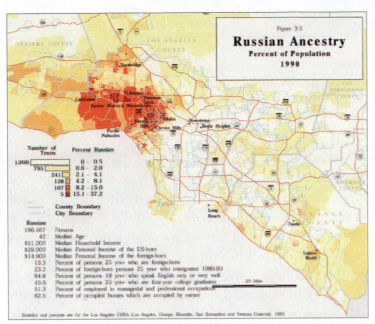

Figure 3.5

Russian Ancestry
Percent of Population
1990

Number of Tracts		Percent Russian
1269		0 - 0.5
795		0.6 - 2.0
241		2.1 - 4.1
128		4.2 - 8.1
107		8.2 - 15.0
5		15.1 - 37.2

County Boundary
City Boundary

Russian
196,467	Persons
42	Median Age
$51,000	Median Household Income
$29,000	Median Personal Income of the US-born
$14,900	Median Personal Income of the foreign-born
13.3	Percent of persons 25 yrs+ who are foreign-born
23.2	Percent of foreign-born persons 25 yrs+ who immigrated 1980-90
94.8	Percent of persons 18 yrs+ who speak English only or very well
45.6	Percent of persons 25 yrs+ who are four-year college graduates
51.3	Percent of employed in managerial and professional occupations
62.5	Percent of occupied homes which are occupied by owner

20 Miles

Statistics and percents are for the Los Angeles CMSA (Los Angeles, Orange, Riverside, San Bernardino and Ventura Counties), 1990.

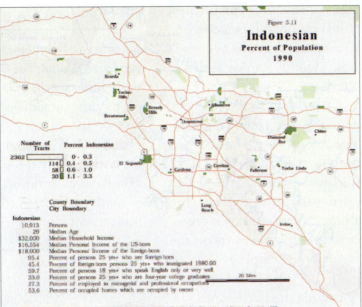

Figure 5.11

Indonesian
Percent of Population
1990

Number of Tracts		Percent Indonesian
2362		0 - 0.3
114		0.4 - 0.5
58		0.6 - 1.0
30		1.1 - 3.3

County Boundary
City Boundary

Indonesian
10,913	Persons
29	Median Age
$32,000	Median Household Income
$16,554	Median Personal Income of the US-born
$18,000	Median Personal Income of the foreign-born
95.4	Percent of persons 25 yrs+ who are foreign-born
45.4	Percent of foreign-born persons 25 yrs+ who immigrated 1980-90
59.7	Percent of persons 18 yrs+ who speak English only or very well
33.0	Percent of persons 25 yrs+ who are four-year college graduates
27.3	Percent of employed in managerial and professional occupations
53.6	Percent of occupied homes which are occupied by owner

20 Miles

Statistics and percents are for the Los Angeles CMSA (Los Angeles, Orange, Riverside, San Bernardino and Ventura Counties), 1990.

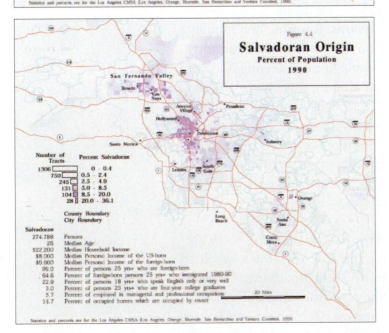

Figure 4.4

Salvadoran Origin
Percent of Population
1990

Number of Tracts		Percent Salvadoran
1306		0 - 0.4
750		0.5 - 2.4
245		2.5 - 4.9
131		5.0 - 8.5
104		8.5 - 20.0
28		20.0 - 36.1

County Boundary
City Boundary

Salvadoran
274,788	Persons
26	Median Age
$22,200	Median Household Income
$8,000	Median Personal Income of the US-born
$9,600	Median Personal Income of the foreign-born
99.0	Percent of persons 25 yrs+ who are foreign-born
64.6	Percent of foreign-born persons 25 yrs+ who immigrated 1980-90
22.9	Percent of persons 18 yrs+ who speak English only or very well
3.0	Percent of persons 25 yrs+ who are four-year college graduates
5.7	Percent of employed in managerial and professional occupations
14.7	Percent of occupied homes which are occupied by owner

20 Miles

Statistics and percents are for the Los Angeles CMSA (Los Angeles, Orange, Riverside, San Bernardino and Ventura Counties), 1990.

Percentiles allowed Allen and Turner to highlight the **outliers**, the extreme high and low values of the data set, choosing either the bottom 10th or 20th percentile, and adding the top 80th, 90th, and 95th percentiles.

Question 8: Study the Russian Ancestry map by Allen and Turner, and compare it to your equal intervals Russian Ancestry map. Both maps show the same data theme. How are they different?

Question 9: Which classification scheme do you think best represents Russian Ancestry in this case, equal intervals or percentiles? Why?

Question 10: Now compare the three maps by Allen and Turner. How do the percentiles of the three ethnic groups compare?

Question 11: In addition to the choropleth, what technique do they use to give the reader an indication of the total distribution of the data set?

Question 12: Is one of the maps a more effective portrayal of the spatial distribution of the outliers, or do all three effectively portray outliers equally? Explain your interpretation.

◐ 5.3
Ethnic Diversity as Cultural Perception

The extent to which a map reflects the ethnic diversity of a region depends on many factors. Diversity in the map depends on who the mapmaker chooses to portray in the map, and the message that the mapmaker hopes to convey to the reader.

Diversity in a map also reflects the mapmaker's perceptions or familiarity with the people in the region. Often, our ability to perceive ethnic diversity depends on how familiar we are with a place; we tend to perceive more ethnic diversity among people whom we know, and less diversity among people whom we don't know. Strange or distant populations often fall into broader categories, or under the anonymous category of "Other," or may not be shown at all. By comparing maps of the same region made by different mapmakers, we can explore changes in how cultures perceive the ethnic diversity around them.

The map to the right is an English copy of a Native map drawn on deerskin about 1721 by an unknown person, as a gift to the new governor of South Carolina, Francis Nicholson. The map delineates Nicholson's new territory for him, showing the connections of the Indian nations in relationship to the southern English colonies.

At the heart of these connected nations was a confederacy of thirteen distinct cultures having separate languages and customs, dominated by the Nasaws. They maintained social and political ties with their neighbors, the Cherokees and Chickasaws, though these latter nations were not part of the confederacy. The English, on the other hand, perceived these separate nations as a single ethnic identity, known as the Catawbas.

The native map is drawn in traditional southeastern native cartographic style, using circles to depict native communities and squares to depict European communities. Charleston is represented as a grid of streets on the left side of the map.

Connecting the circles and squares are double line connections. Running horizontally through the map is a prominent double line connecting Nasaw to Charleston; this line is labeled "The English Path to Nasaw." Scholars have speculated that the lines represent both rivers and paths between native peoples, as well as social and political connections.

As you study the map, you will see that the tribes are symbolized by different circle sizes. Curious about the meaning of the different sizes, anthropologist Gregory Waselkov compared the content of the map to population records from the colonial period, to see whether the circles were similar to the Western technique of using proportional symbols to show population data. He found that there was some correlation between native populations and circle sizes, but not for all of the circles.

Question 1: What other factors besides population size do you think might influence the circle sizes showing the different nations?

Question 2: What do you think were the motivations of the native author or authors for portraying this level of cultural complexity in a map for a new colonial governor?

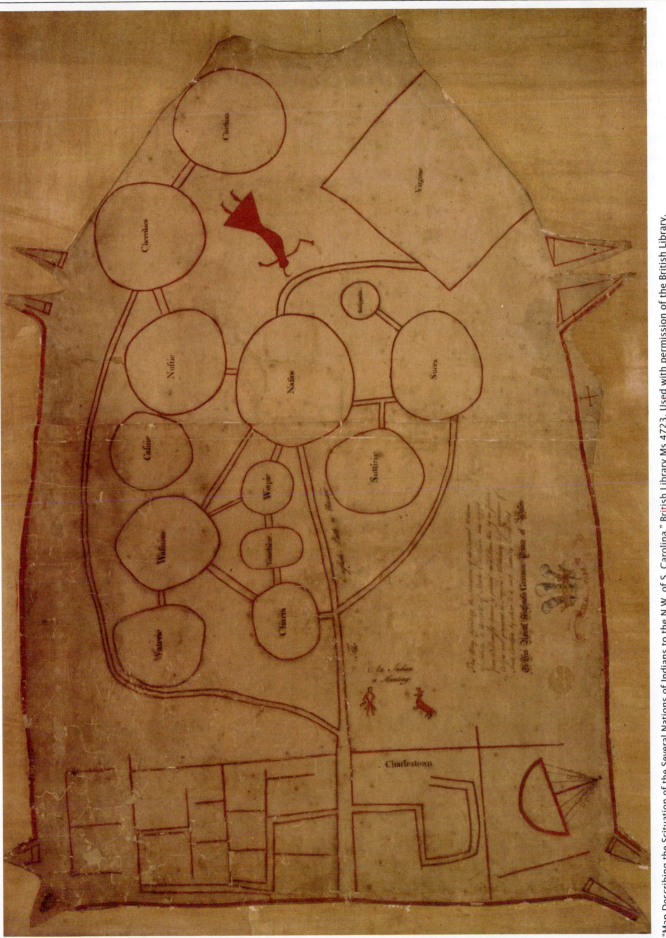

"Map Describing the Scituation of the Several Nations of Indians to the N.W. of S. Carolina," British Library Ms 4723. Used with permission of the British Library.

A second map of the southeastern region shows a very different view of diversity in the region. The map on the right is from the French royal geographer Guillaume de l'Isle and depicts the colony of Carolina in 1718. At that time, the French and English were competing with each other for both territorial claims and Indian allegiances in the Southeast.

Study the place name details in de l'Isle's map. To get oriented, Charleston, South Carolina, is shown as "Charles Town" on the map. When you turn the page sideways to study the place names, Charles Town is on the coast in the lower-right-hand corner of the map.

Question 3: Compare the de l'Isle map to the "Catawba" map on page 65. How do the two maps compare in terms of the way that they depict the English colonial claims?

Question 4: How do the two maps differ in the symbols used to differentiate Indian nations? How do they differ in their portrayal of native towns?

Question 5: For whom do you think de l'Isle's map was created? How do you think this influenced his portrayal of ethnicity in the map?

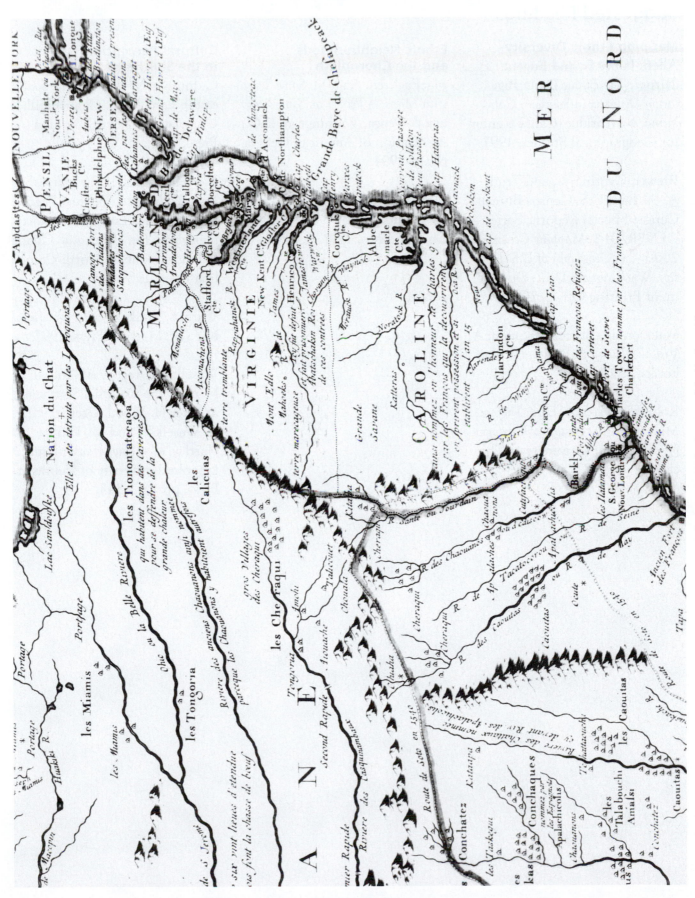

Sources and Suggested Readings

Mapping Ethnic Diversity

Allen, James P., and Eugene Turner. *The Ethnic Quilt: Population Diversity in Southern California*. Northridge, Calif.: Center for Geographical Studies, 1997.

Brewer, Cynthia A., and Trudy A. Suchan, U.S. Census Bureau, Census Special Reports, Series CENSR/01-1. *Mapping Census 2000: The Geography of U.S. Diversity*. Washington, D.C.: Government Printing Office, 2001.

MacDonald, Bruce. *Vancouver: A Visual History*. Vancouver: Talonbooks, 1994.

Robinson, Arthur. *Early Thematic Mapping in the History of Cartography*. Chicago: University of Chicago Press, 1982.

Ethnic Neighborhoods and the Choropleth

MacEachren, Alan M. *Some Truth with Maps: A Primer on Symbolization & Design*. Washington, D.C.: Association of American Geographers, 1994.

Monmonier, Mark. *How to Lie with Maps*. Chicago: University of Chicago Press, 1991.

Cultural Perception in the Southeast

Cumming, William P. *The Southeast in Early Maps*. Chapel Hill: University of North Carolina Press, 1962.

Merrell, James H. *The Indians' New World: Catawbas and Their Neighbors from European Contact through the Era of Removal*. Chapel Hill: University of North Carolina Press, 1989.

Warhus, Mark. *Another America*. N.Y.: St. Martin's Press, 1997.

Waselkov, Gregory A. "Indian Maps of the Colonial Southeast," in Peter H. Wood, Gregory A. Waselkov, and M. Thomas Hartley, eds. *Powhatan's Mantle*. Lincoln: University of Nebraska Press, pp. 292–343.

CHAPTER

6

Politics

✪ 6.1
The Politics of Projections

The decision about whose political worldview will be represented in a map has inspired some of the most heated debates in geography and politics. Of these debates, perhaps the most contentious is the question of which map projection is the best for showing the countries of the world. This seemingly simple task of choosing a projection that shows all countries with the minimum of distortion has been one of the most difficult and devisive challenges for political geographers and cartographers to solve.

Why is this such a difficult task to accomplish? One reason has to do with the way that lines of latitude (**parallels**)

and lines of longitude (**meridians**)

are arranged when they are removed from the spherical earth.

On the globe, parallels and meridians form a tidy grid called the **graticule:**

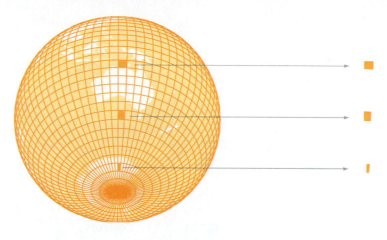

Base map data source: ESRI 1999.

Parallels form concentric circles along the surface of the globe. Meridians, on the other hand, converge gradually toward a point at each pole. The resulting cells of the graticule vary in shape and size, depending on where they fall on the globe. In other words, the cells of the grid on the round globe don't resemble the cells of a grid on flat paper. Before a graticule is flattened, the individual cells already have different sizes and shapes.

Flattening these round cells requires a considerable amount of readjustment to the shapes and sizes of its quadrilaterals. One small readjustment to the graticule will create distortion in the way that the outlines of countries are depicted on the earth. As the graticule is progressively flattened out, these distortions mul-

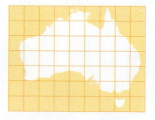

tiply. This arrangement and flattening of the graticule is a **map projection.**

To control the way that a map projection distorts the earth, the cartographer decides which properties of the graticule are the most important to preserve. These properties are: **area, distance, direction,** and **shape.** Is the area of a country the most important factor to preserve? If so, then distance and shape are going to suffer. If shape is the most important factor, then the areas of the countries will be distorted.

All properties cannot be preserved in a map projection; some part of the properties of the graticule have to be sacrificed. Which properties are preserved depends on the way in which the graticule is "peeled" from the globe. In general, there are three ways to peel and project the graticule.

A **cylindrical projection** is created by projecting the graticule onto a plane wrapped around the globe in a cylinder:

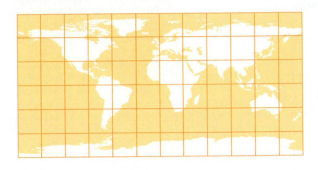

An **azimuthal** or **planar projection** is created by projecting the graticule from a point on the globe onto a flat plane:

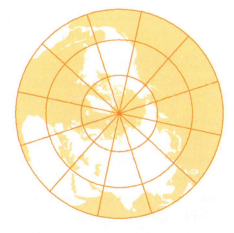
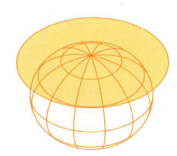

A **conic projection** is created by projecting the graticule onto a plane wrapped around the globe in a cone:

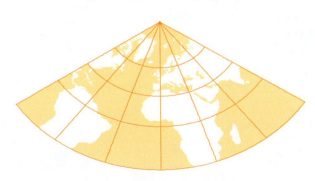
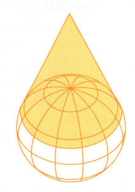

Base map data sources: ESRI 1999.

Most projections in use today begin as cylindrical, azimuthal, or conic shapes, and are mathematically modified to better preserve properties of the graticule.

The Search for a Political View of the World

There are hundreds of map projections available for us to use, each suited for different purposes depending on their qualities. Geographers portraying the nations of the Western hemisphere often choose the Transverse Mercator Projection, for example. The nations of the African continent are often depicted by the Azimuthal Equal-Area Projection. For the whole earth, physical geographers sometimes favor Goode's World Projection because of its ability to depict the connectivity of landforms on the earth's surface.

When it comes to depicting global political geography, however, the process becomes tougher. The choice of projection for showing the nations of the world and their relationship to each other is one of the most heated cartographic debates to come to the public's attention.

The source of the debate stems from two projections. Perhaps the most well-known projection is the Mercator, a cylindrical projection originally developed for navigational purposes by Gerhard Mercator in 1569.

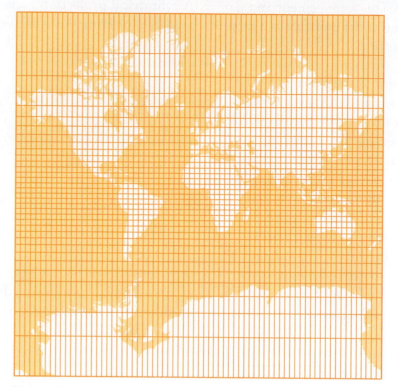

Mercator Projection. Base map data source: ESRI 1999.

The Mercator projection preserves the graticule property of direction while forsaking area and shape. In part because it is easy to make, the Mercator has become a popularly reprinted wallmap for schools. Its utility for educational purposes, however, is dubious, because maintaining perfect direction creates significant distortion of shape and area. In the equatorial region, the distortion is minimal. In the high latitudes, however, the shape and size of landmasses are significantly distorted.

Another projection that became widespread during the twentieth century was the Van der Grinten, the official projection of the National Geographic Society. The Van der Grinten also depicted large amounts of distortion in the high latitudes.

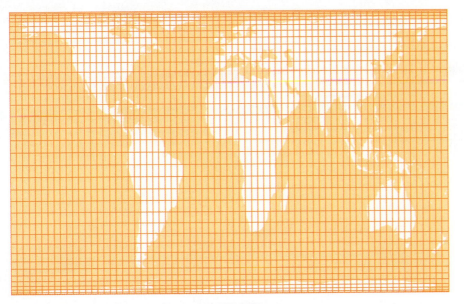

Gall-Peters Projection. Base map data source: ESRI 1999.

The dismal popular choices for looking at world political geography bothered Arno Peters, for whom the distortions comprised a social injustice of misrepresentation to the marginalized countries of the world. To Peters, political geography could be served only by preserving the property of area above all other properties of the graticule.

In 1972, dissatisfied with the conventional displays of distortions, he developed what he called the Peters Projection (now referred to as the Gall-Peters projection because of its derivation from a map by James Gall in 1855). The Peters map depicted the countries of the world in proportionate area in order to better represent developing countries in relationship to the developed world.

Question 1: Study the coastlines as they are depicted in the Mercator projection, and compare these to the Gall-Peters projection. Where are the maps most different?

Question 2: Is there any region or country that appears the same in both the Mercator and the Gall-Peters? Why is that?

Question 3: Do you agree that area is the most important criterion for comparing the spatial relationship between countries? What other characteristics of a political body besides physical size might be useful for making comparisons?

Peters' map was embraced by relief organizations such as UNICEF and the World Health Organization, in which there was great interest in fair cartographic representation for developing nations. Today, it continues to be a popular wallmap.

Cartographers, on the other hand, have reacted negatively to Peters' map. For some, Peters' map is aesthetically offensive. Others have found it unacceptable as a world map because it preserves only one of the graticule's properties, an extreme solution for a map of the whole earth. This poor academic reception aside, Peters' map revitalized the cartographic debate over best fit solutions for projections of the political world. Because area, distance, direction, and shape are all important factors in the way a country is depicted, many cartographers now advocate that, for a general world political map, there can only be a compromise.

Winkel Tripel Projection. Base map data source: ESRI 1999.

Compromise, Compromise

After Peters, work on development of a better all-purpose world map projection was revitalized, and cartographers called for a rethinking of the benefits of the **compromise projection.**

A compromise projection preserves none of the properties of the graticule. Instead, it minimally distorts all properties in order to achieve a balance, or compromise, of distortions across the map. The current projection advocated by the National Geographic Society, the **Winkel Tripel,** is just such a compromise projection.

The Winkel Tripel is a modified azimuthal projection created by Oswald Winkel in 1921 as a middle ground between two other projections, the Equidis-
tant Cylindrical and the Aitoff. In 1998, the National Geographic Society adopted it as the new standard for their world maps. The Winkel Tripel minimizes the distortion of shape, area, and direction, except in the very high latitudes at the poles. Just as Mercator's projection met the need for world maps in the sixteenth century to depict the earth in a way that would preserve the property of direction above all other graticule properties, the Winkel Tripel meets the current need for a world map to depict all of the earth except the polar regions by balancing shape, area, and direction distortions as evenly as possible.

Question 4: Compare the coastlines of the Winkel Tripel to a globe in your classroom. What compromises have been made in this map?

Question 5: What is the most important property of a country to preserve in a map?

Question 6: What will be the priorities of world projections twenty years from now, as globalization intensifies?

Question 7: To what degree should aesthetics play a role in the creation of a political world map?

6.2 Evaluating Redistricting Online

Another aspect of political geography in which mapping plays a primary role is **redistricting.** Redistricting is the process of creating new electoral district boundaries, which in turn creates new **electoral regions.**

As you have learned in class, new electoral district boundaries may be illegally manipulated or **gerrymandered** to favor certain sectors of society, such as political parties, ethnic groups, or age groups.

According to U.S. law, all states and counties with more than one electoral district must redistrict every ten years, to reflect new population figures released by the U.S. Census. Finding a way to redistrict so as to represent every citizen fairly, without gerrymandering, is a complex task of analyzing the geography of race, age, and political persuasion in a given district.

In this exercise we will analyze electoral redistricting in Hillsborough County, Florida, and compare the redistricting maps to the demographic characteristics of the county.

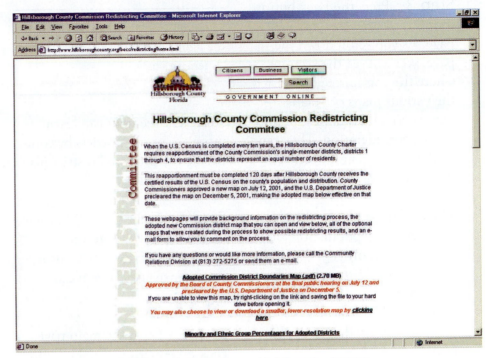

Screen shot courtesy of Hillsborough County Geographic Information Systems.

Step 1 Launch your browser and go to the home page of the Hillsborough County Commission Redistricting Page at **www.hillsboroughcounty.org/ bocc/redistricting/home.html.**

You should see a gateway to the Web site as in the screen shot, above. This Web site describes the redistricting process that Hillsborough County underwent in 2001 for the reapportionment of its County Commission.

Step 2 Scroll down and click on the link "Criteria for Redistricting of County Commission."

Question 1: Briefly summarize the main criteria that Hillsborough County considered in its redistricting process.

Step 3 Press the Back button to return to the previous Web page. As you scroll down the page, you will see the sections where the committee has posted the various maps or "plans" of new district boundaries submitted for consideration. They can be downloaded and opened as PDF files using Adobe Acrobat.

Three of these plans, with their corresponding population data tables, appear on page 77. In the maps, the old district boundaries are depicted as red and white lines, and the proposed new districts are depicted as shaded areas of different colors.

Question 2: Compare the old district boundaries in each of the three maps, B, D, and G. Which district is the most compact? What geographical factors (physical or human) might have influenced the shape of this district?

Question 3: Which of the former electoral districts is the least compact? What geographical factors might have influenced this shape?

Question 4: Compare the shapes of the proposed district boundaries in each of the three maps. Which of the three plans shows the greatest compactness?

Question 5: Evaluate the data on race in the tables to the right of the maps. Is there one plan that particularly favors the Hispanic vote? Why or why not?

Question 6: How would you evaluate the plans in terms of the African–American vote?

Next, let's see if we can learn more about the districts by comparing the proposed district plans to census data.

Step 4 Leave the Hillsborough County site and go to the home page of the U.S. Census, **www.census.gov,** as you did in chapter 5.

Step 5 From the main page, click on American FactFinder in the left column. From the Factfinder page, scroll down to the Maps section and click on "Thematic Maps."

This launches the interactive map application of FactFinder.

Step 6 From the horizontal menu bar at the top of the page, go to "Change Selections" and choose "Geography."

Step 7 From the next window, change "Select a Geographic Type" to "County," change "Select a State" to "Florida," and change "Select a Geographic Area" to "Hillsborough County."

Step 8 Click "Show Map."

You should now have a map view showing "Persons per Square Mile" in Hillsborough County, as in the screen shot below.

Screen shot from U.S. Census Bureau, American FactFinder (http://factfinder.census.gov), 2001.

You already have information about race for each of the districts. Are there other socio-economic factors that might be considered in the districts? Below, we will explore two other demographic variables, age and income, to see if there is a correlation.

Step 9 Go to the horizontal menu bar at the top of the page, and under "Change Selections" choose "Theme."

Step 10 From the list of available themes, scroll down and select "Percent of Persons 65 Years and Over: 2000," and click "Show Map."

Step 11 From the Zoom options to the right of the map, click on button 6 to zoom in one level.

Question 7: What is the enumeration unit of this map?

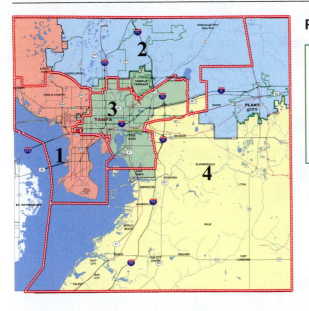

Plan B

	Total Population	White*	African-American	Native American	Asian-American	Other	Hispanic**	Non-Hispanic
DISTRICT 1	249,775	204,263 (82%)	17,893 (7%)	829 (0.3%)	7,906 (3%)	18,690 (7%)	56,383 (23%)	193,392 (77%)
DISTRICT 2	249,708	200,210 (80%)	25,115 (10%)	856 (0.3%)	6,901 (3%)	16,477 (7%)	36,278 (15%)	213,430 (85%)
DISTRICT 3	249,777	133,700 (54%)	91,107 (36%)	1,139 (0.5%)	3,453 (1%)	20,161 (8%)	52,950 (21%)	197,197 (79%)
DISTRICT 4	249,688	212,730 (85%)	15,308 (6%)	1,055 (0.4%)	3,697 (1%)	10,741 (7%)	34,081 (14%)	215,607 (86%)
Total	998,948							

* Numbers shown in parenthesis represent percentage of racial classification by district. Percentages may not equal 100% when added due to rounding.

** Americans of Hispanic or Latino decent are considered members of a language or ethnic minority.

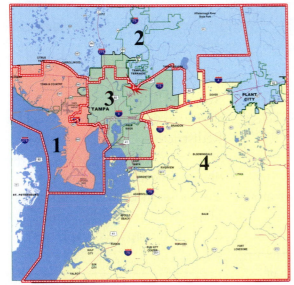

Plan D

	Total Population	White*	African-American	Native American	Asian-American	Other	Hispanic**	Non-Hispanic
DISTRICT 1	249,106	199,421 (80%)	21,041 (8%)	833 (0.3%)	7,686 (3%)	20,125 (8%)	64,407 (26%)	184,699 (74%)
DISTRICT 2	250,026	209,676 (84%)	18,821 (8%)	836 (0.3%)	6,976 (3%)	13,717 (5%)	31,478 (13%)	218,548 (87%)
DISTRICT 3	249,976	129,439 (52%)	94,923 (38%)	1,175 (0.5%)	3,640 (1%)	20,799 (8%)	47,000 (19%)	202,976 (81%)
DISTRICT 4	249,840	212,367 (85%)	14,638 (6%)	1,035 (0.4%)	3,645 (1%)	18,155 (7%)	36,807 (15%)	213,033 (85%)
Total	998,948							

* Numbers shown in parenthesis represent percentage of racial classification by district. Percentages may not equal 100% when added due to rounding.

** Americans of Hispanic or Latino decent are considered members of a language or ethnic minority.

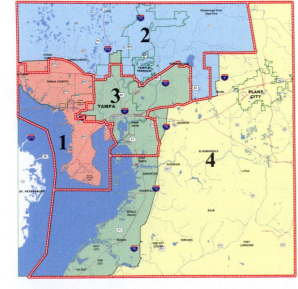

Plan G

	Total Population	White*	African-American	Native American	Asian-American	Other	Hispanic**	Non-Hispanic
DISTRICT 1	247,212	199,492 (81%)	19,239 (8%)	831 (0.3%)	7,685 (3%)	19,965 (8%)	62,764 (25%)	184,448 (75%)
DISTRICT 2	249,441	249,441 (83%)	20,126 (8%)	875 (0.4%)	8,013 (3%)	12,915 (5%)	32,399 (13%)	217,042 (87%)
DISTRICT 3	249,069	132,763 (53%)	91,583 (37%)	1,194 (0.5%)	2,580 (1%)	20,949 (8%)	48,614 (20%)	200,455 (80%)
DISTRICT 4	253,226	211,136 (83%)	18,475 (7%)	979 (0.4%)	3,669 (1%)	18,967 (7%)	35,915 (14%)	217,311 (86%)
Total	998,948							

* Numbers shown in parenthesis represent percentage of racial classification by district. Percentages may not equal 100% when added due to rounding.

** Americans of Hispanic or Latino decent are considered members of a language or ethnic minority.

Plans B, D, and G courtesy of Hillsborough County Geographic Information Systems.

Step 12   Because we want to look at the most detailed information about race that we can, change the enumeration unit to Block Group.

In U.S. Census data, a **block group** is one of the smallest enumeration units available, averaging about 1,500 people in size.

You should now have a view of the redistricting area showing percent 65 and over by block group, as in the screen shot at the right.

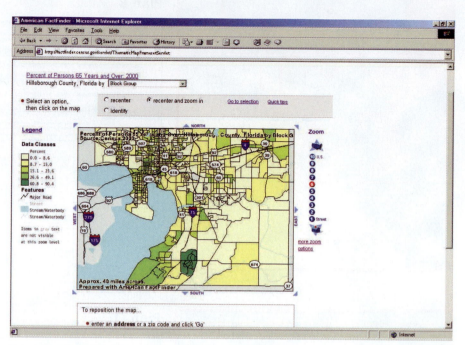

Screen shot from U.S. Census Bureau Web site, American FactFinder (http:/factfinder.census.gov), 2001.

Question 8: Compare the Census map with the proposed district boundary maps. (Remember that you can use the zoom and pan tools to explore different regions of the county.) Are there any districts that appear to favor people over the age of 65? If so, explain.

Step 13   Go to the "Change Selections" menu bar, but this time choose "Data Set."

Step 14   Select Summary Tape File 3 and click "Next."

Step 15   As before, for "Select a Geographic Type," choose "County;" for "Select a State," choose "Florida;" and for "Select a Geographic Area," choose "Hillsborough County."

Step 16   Click "Next."

Step 17   From the list of income-related variables, select "Per Capita Income 1989: 1990" and click "Show Map." Be patient—this may take a few moments to appear on your monitor.

Question 9: Compare the map of per capita income to the proposed district boundary plans. Are there any districts that appear to favor people with higher incomes? Explain your reasoning.

Question 10: Based on what you have observed so far, which of the three proposed plans do you think appears to be the most equitable redistricting solution, and why?

Step 18   Return to the Hillsborough County Commission Redistricting Committee Web page as in Step 1, and click on the link to the "Adopted Commission District Boundaries Map" (you can use the lower-resolution map link if you like).

Question 11: Do you think the Committee found an equitable redistricting solution? Consider in your answer the criteria that the committee set forth, from Question 1, above.

6.3
Political Boundaries, Generalization, and Scale

When we read a map for information about political boundaries and borders, we are dependent on the mapmaker's skill with **generalization.** Generalization is simply the term used by cartographers for the process of deciding how much geographical detail is going to be shown in a map. All maps are generalizations, because all maps are scaled-down versions of reality.

The level at which a political boundary is generalized on a map depends on the scale of the map (How much room is there to show these boundary details?), the quality of the original data (How much detail was known about the boundary to begin with?), and the intentions of the cartographer (Whose political claim is being shown?).

The balance that is struck between these factors can be a precarious one. The omission of a detail about one curve in an international border may result in inaccurate decision making, or cause grave offense for one or both countries.

The border between Eritrea and Ethiopia is a good example of how cartographic generalization can affect the the relations between countries. Until 1952, Eritrea was a colony of Italy. Eritrea became independent from Italy following the Second World War, and in 1962, Ethiopia annexed the region as its new, northern province.

The map below is from a world atlas published in 1952, when Eritrea was on the eve of independence from Italy.

Question 1: Compare the difference in boundary-line symbolization between Ethiopia's northern border with Eritrea and its western border with Sudan. How has the cartographer differentiated these two kinds of boundaries?

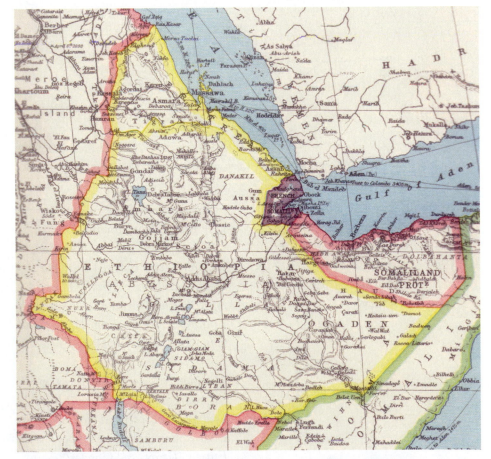

Detail from "Africa, North-East," in *The Citizen's Atlas of the World.* 10th ed. © Bartholomew Ltd 1952 Reproduced by Kind Permission of HarperCollins Publishers www.bartholomewmaps.com.

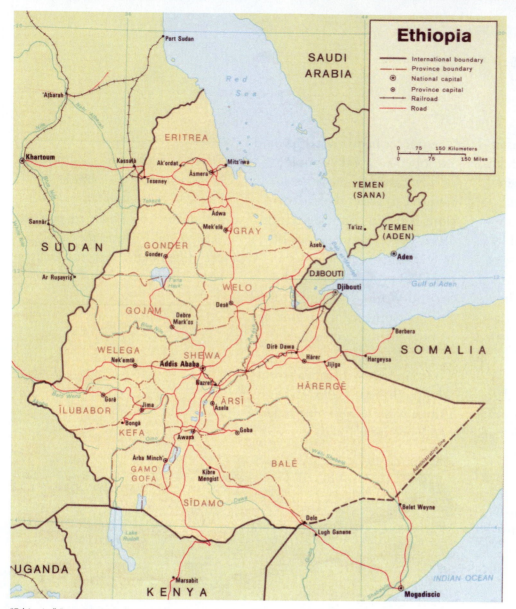

"Ethiopia." Base 504020, Central Intelligence Agency, Government Printing Office, January 1979.

In the map above, showing the same region in 1979, Ethiopia has been transformed by the 1962 annexation of Eritrea. The old international border is now a **relic boundary.**

Question 2: Compare the 1951 and 1979 maps of Ethiopia. How has the annexation changed Ethiopia's geographical context in the Horn of Africa? What advantages or disadvantages can you see for the people of Ethiopia? For the people of Eritrea?

Question 3: Can you find any indication in the 1979 map that there used to be an international border at what is now the provincial border of Eritrea, Tigray, and Gonder?

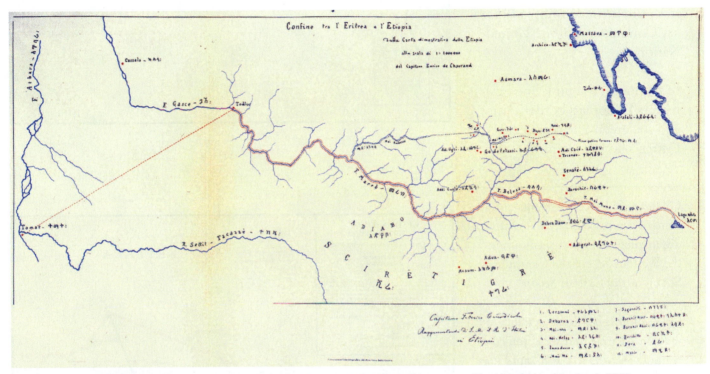

1900 Treaty Map, Eritrea-Ethiopia Boundary Commission, United Nations. EEBC Web site (www.un.org./NewLinks/eebcarbitration/), 2002.

Generalization and Interpretation

In 1993, Eritrea asserted its independence from Ethiopia. When this political independence extended to economic autonomy through the issuing of Eritrean currency in 1998, war broke out between the two countries.

The main dispute in the war centered on the delineation of the Eritrean-Ethiopian international border. The border was based on the old colonial boundaries of Eritrea as set by a 1900 Italian treaty, shown above. In the treaty, the border was described and graphically portrayed with a high degree of generalization, with the understanding that a more detailed map would be produced later.

Question 4: Study the treaty map above. How is the colonial boundary depicted? Is this a natural, ethnographic, or geometric boundary?

Despite its provisional nature, a less generalized map of the border was never produced, and the treaty map remained the primary description for the border. As a result, each country interpreted the specifics of the boundary location differently.

The conflicting interpretations are illustrated in these three maps from the Eritrea-Ethiopia Boundary Commission of the United Nations.

In the maps, the pink line represents Ethiopia's border claim, and the green line represents Eritrea's border claim.

Question 5: Which segments of the border seem to be under the greatest dispute between the two countries: the Western Sector, the Central Sector, or the Eastern Sector?

Question 6: Look back to the treaty map on page 81 and see if you can relate the sectors to the colonial illustration. Which aspects of the treaty map seems to be causing the greatest conflict?

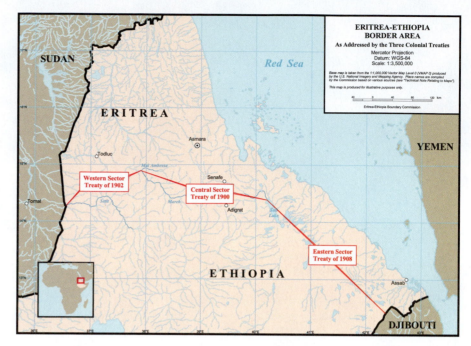

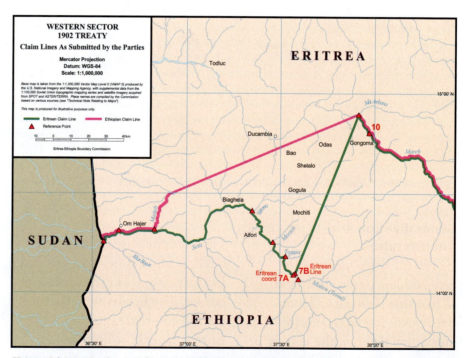

"Eritrea-Ethiopia Border Area," top, and "Western Sector 1902 Treaty," bottom, Eritrea-Ethiopia Boundary Commission, United Nations. EEBC Web site (www.un.org./NewLinks/eebcarbitration/), 2002.

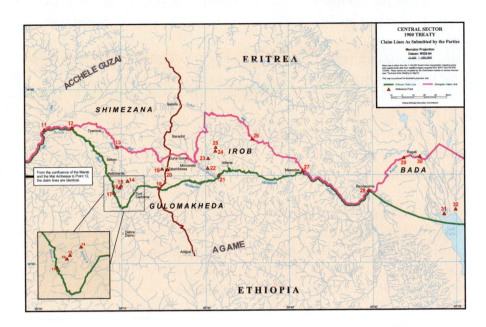

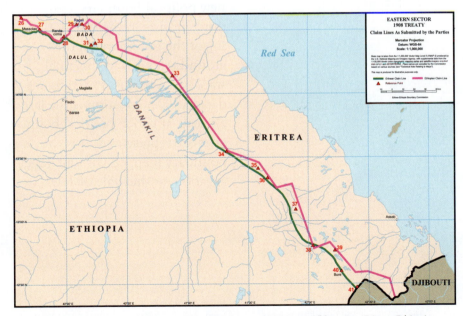

"Central Sector 1902 Treaty," top, and "Eastern Sector 1902 Treaty," bottom, Eritrea-Ethiopia Boundary Commission, United Nations. EEBC Web site (www.un.org./NewLinks/eebcarbitration/), 2002.

After 80,000 lives were lost in this conflict, a peace agreement between the two countries was finally forged in December 2000, and a commission was established to make an objective ruling on the border. The Eritrea-Ethiopia Boundary Commission studied the treaty map and border description carefully, and in April 2002, released a more specific description of the border, divided according to the three sectors of the previous maps.

In Search of a New Border

For the last part of this exercise, imagine that you are in the boundary drawing seat of the commission, implementing the new boundary as mandated by the United Nations.

Step 1   Photocopy the three blank base maps, for each of the three sectors, on pages 87–89.

Step 2   For each sector, read the new description of the boundary as handed down by the commission, and plot the new delineation of the Eritrea- Ethiopia border on the corresponding base map of the region.

Question 7: Analyze the three new sector maps you have created, and compare them to the old boundary map. Is this new international boundary more geometric? More natural?

Question 8: Will there be a new relic political boundary? If so, where?

Question 9: Compare the new boundary to the previous claims of Eritrea and Ethiopia. Where did Eritrea make concessions to Ethiopia? Did Ethiopia in turn make concessions to Etritrea?

You will notice that there are several turning points from the border description that remain vague. The final stage of setting a new political boundary, demarcation, consists in the actual marking of the boundary on the ground, and it is in this final stage when these last details are decided. Demarcation is the largest scale at which political boundaries are represented. It is the boundary mapped without generalization, at the scale of 1:1.

Decision

For the reasons set out above, the Commission unanimously decides that the line of the boundary between Eritrea and Ethiopia is as follows:

A. In the Western Sector

(i) The boundary begins at the tripoint between Eritrea, Ethiopia and the Sudan and then runs into the centre of the Setit opposite that point (Point 1).

(ii) The boundary then follows the Setit eastwards to its confluence with the Tomsa (Point 6).

(iii) At that point, the boundary turns to the northeast and runs in a straight line to the confluence of the Mareb and the Mai Ambessa (Point 9).

B. In the Central Sector

(i) The boundary begins at the confluence of the Mareb and the Mai Ambessa (Point 9).

(ii) It follows the Mareb eastwards to its confluence with the Belesa (Point 11).

(iii) Thence it runs upstream the Belesa to the point where the Belesa is joined by the Belesa A and the Belesa B (Point 12).

(iv) To the east and southeast of Point 12, the boundary ascends the Belesa B, diverging from that river so as to leave Tserona and its environs to Eritrea. The boundary runs round Tserona at a distance of approximately one kilometre from its current outer edge, in a manner to be determined more precisely during demarcation.

(v) Thereafter, upon rejoining the Belesa B, the boundary continues southward up that river to Point 14, where it turns to the southwest to pass up the unnamed tributary flowing from that direction, to the source of that tributary at Point 15. From that point it crosses the watershed by a straight line to the source of a tributary of the Belesa A at Point 16 and passes down that tributary to its confluence with the Belesa A at Point 17. It then continues up the Belesa A to follow the Eritrean claim line to Point 18 so as to leave Ford Cadorna and its environs within Eritrea.... Point 18 lies 100 metres west of the centre of the road running from Adigrat to Zalambessa.

(vi) From Point 18, the boundary runs parallel to the road at a distance of 100 metres from its centre along its western side and in the direction of Zalambessa until about one kilometre south of the current outer edge of the town. In order to leave that town and its environs to Ethiopia, the boundary turns to the northwest to pass around Zalambessa at a distance of approximately one kilometre from its current outer edge until the boundary rejoins the Treaty line at approximately Point 20, but leaving the location of the former Eritrean customs post within Eritrea. The current outer edge of Zalambessa will be determined more precisely during the demarcation.

(vii) From Point 20 the boundary passes down the Muna until it meets the Enda Dashim at Point 21.

(viii) At Point 21 the boundary turns to the northwest to follow the Enda Dashim upstream to Point 22. There the boundary leaves that river to pass northwards along one of its tributaries to Point 23. There the boundary turns northeastwards to follow a higher tributary to its source at Point 24.

(ix) At Point 24 the boundary passes in a straight line overland to Point 25, the source of one of the headwaters of a tributary of the Endeli, whence it continues along that tributary to Point 26, where it joins the Endeli.

(x) From Point 26, the boundary descends the Endeli to its confluence with the Muna at Point 27.

(xi) From Point 27, the boundary follows the Muna/Endeli downstream. Near Rendacoma, at approximately Point 28, the river begins also to be called the Ragali.

(xii) From Point 28, the line continues down the Muna/Endeli/Ragali to Point 29, northwest of the Salt Lake, and thence by straight lines to Points 30 and 31, at which last point this sector of the boundary terminates.

C. In the Eastern Sector

The boundary begins at Point 31 and then continues by a series of straight lines connecting ten points, Points 32 and 41. Point 41 will be at the boundary with Djibouti. Point 40, lies equidistantly between the two checkpoints at Bure.

> — Eritrea-Ethiopia Boundary Commission,
> *Decision Regarding Delimitation of the Border between The State of Eritrea and The Federal Democratic Republic of Ethiopia,* April 13, 2002.

Maps on pp. 87-89 ("Western Sector," "Central Sector," and "Eastern Sector") adapted from Eritrea-Ethiopia Boundary Commission, United Nations. EEBC Web site (www.un.org./NewLinks/eebcarbitration/), 2002.

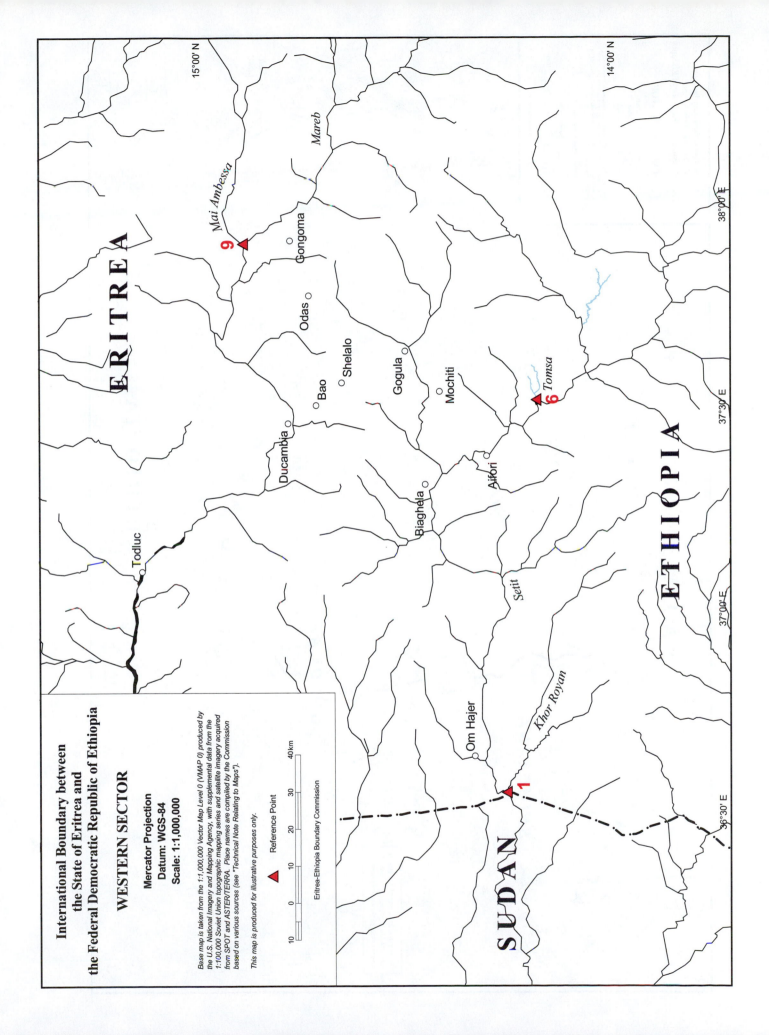

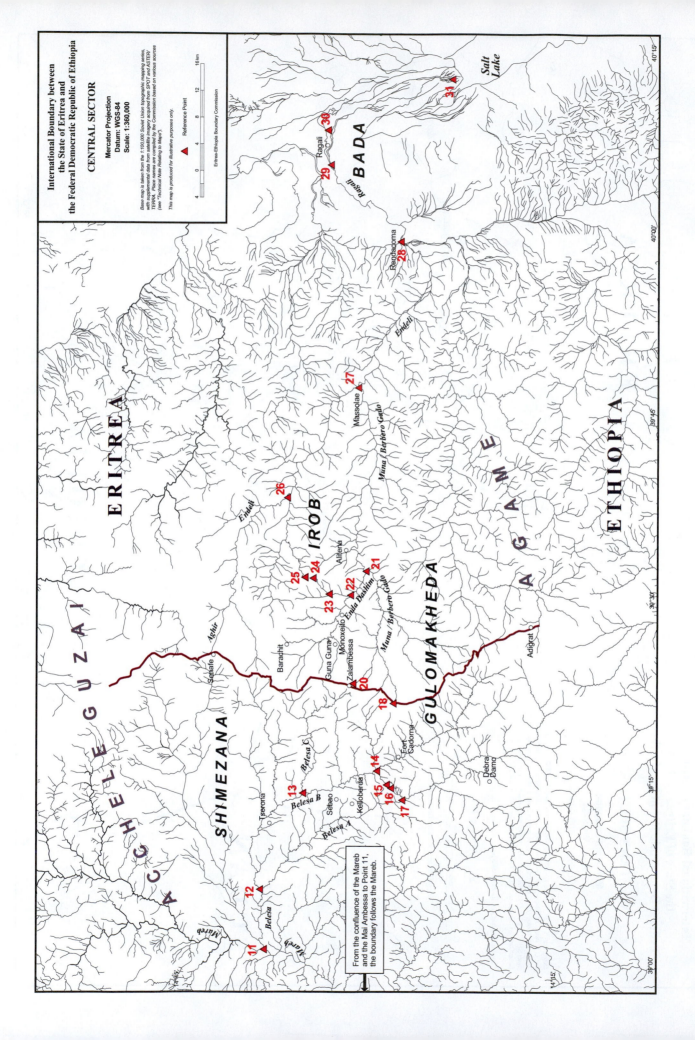

International Boundary between
the State of Eritrea and
the Federal Democratic Republic of Ethiopia

CENTRAL SECTOR

Mercator Projection
Datum: WGS-84
Scale: 1:360,000

Base map is taken from the 1:100,000 Soviet Union topographic mapping series,
with supplemental data from satellite imagery acquired from SPOT and ASTER/
TERRA. Place names are compiled by the Commission based on various sources
(see "Technical Note Relating to Maps").

This map is produced for illustrative purposes only.

▲ Reference Point

Eritrea-Ethiopia Boundary Commission

0   4   8   12   16 km

ERITREA

ETHIOPIA

BADA

Salt Lake

IROB

AGAME

GULOMAKHEDA

SHIMEZANA

ACCHELE GUZAI

From the confluence of the Mareb
and the Mai Ambessa to Point 11,
the boundary follows the Mareb.

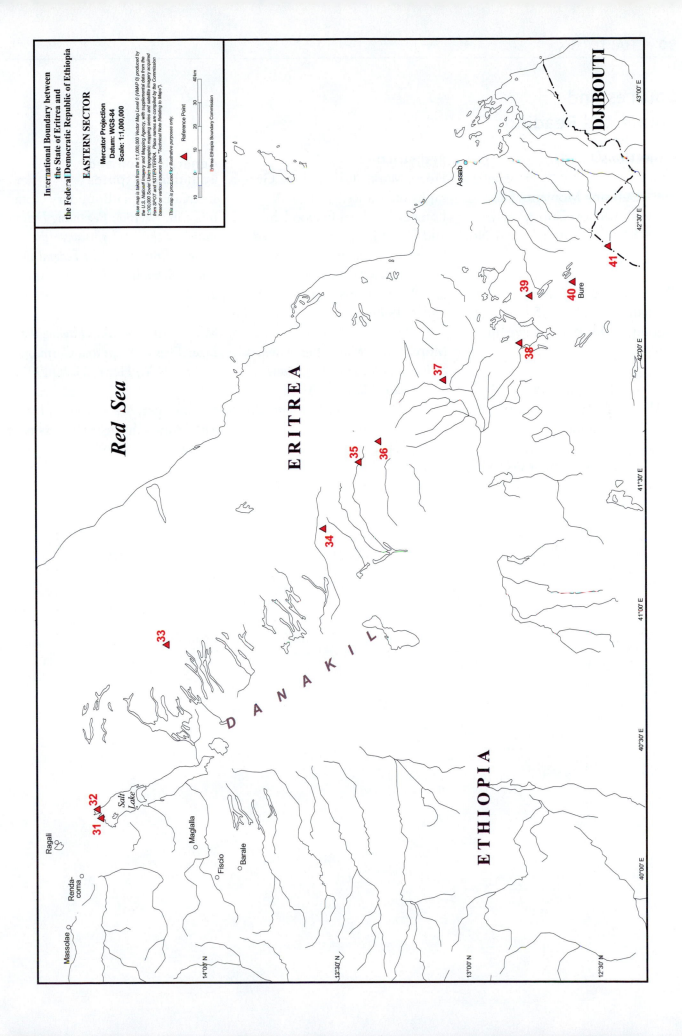

Red Sea

ERITREA

DJIBOUTI

Assab

D A N A K I L

ETHIOPIA

Massolae

Ragali

Renda-
coma

Fiscio

Maglalla

Barale

*Salt
Lake*

Bure

43°00' E
42°30' E
42°00' E
41°30' E
41°00' E
40°30' E
40°00' E

14°00' N
13°30' N
13°00' N
12°30' N

31 32 33 34 35 36 37 38 39 40 41

Sources and Suggested Readings

Projections

Crampton, Jeremy. "Cartography's Defining Moment: The Peters Projection Controversy 1974–1990." *Cartographica* 31 No. 4 (1994):16–32.

Dent, Borden D. *Cartography: Thematic Map Design.* 5th ed. Dubuque: WCB McGraw-Hill, 1999.

Snyder, John P., and Philip M. Voxland. *An Album of Map Projections.* U.S. Geological Survey Professional Paper 1453. Washington, D.C.: Government Printing Office, 1989.

Redistricting

Horn, Mark. "GIS and the Geography of Politics," in Paul A. Longley, Michael F. Goodchild, and David J. Maguire, et al., eds. *Geographical Information Systems.* Volume 2: Management Issues and Applications. 2nd ed. N.Y.: Wiley, 1999.

Monmonier, Mark. *Bushmanders & Bullwinkles: How Politicians Manipulate Electronic Maps and Census Data to Win Elections.* Chicago: The University of Chicago Press, 2001.

Generalization and Boundary Disputes

Eritrea-Ethiopia Boundary Commission. *Decision Regarding Delimitation of the Border between the State of Eritrea and the Federal Democratic Republic of Ethiopia.* April 2002.

Monmonier, Mark. *Drawing the Line: Tales of Maps and Cartocontroversy.* N.Y.: Henry Holt, 1995.

Monmonier, Mark. *How to Lie with Maps.* Chicago: University of Chicago Press, 1991.

7

Population

Vocabulary applied in this chapter
geodemography
population indicators
birth rate
death rate
total fertility rate (TFR)
infant mortality
age distribution
GNPpc

New vocabulary
metadata
number line
natural breaks classification
   scheme
dweller density map
isarithmic map
worker density map

🔍 7.1
Population Indicators I: A Global View

Geodemography is one of the most commonly mapped themes in geography because population data is readily available and lends itself well to mapping, particularly at the global level. Mapping geodemography allows us to go beyond basic population numbers to the **population indicators** that give us a more complex picture of the population dynamics of a place, such as **birth rate, death rate, total fertility rate (TFR)**, and **infant mortality rate.** This exercise gets you started comparing population indicators at a global scale.

One Web site where you can explore geodemography is the United Nations Environment Programme (UNEP)'s "GEO Data Portal." In addition to the numerous environmental databases available here, this site allows you to explore a number of different population indicators.

Step 1 Launch your browser and go to the GEO Data Portal at **http://geodata.grid.unep.ch/**. This will take you to the main page for the GEO Data Portal, shown top, right.

First, let's focus on the Fertility Rate data. As you learned in your textbook, fertility rate is a relatively useful indicator of forthcoming changes in population density for a country.

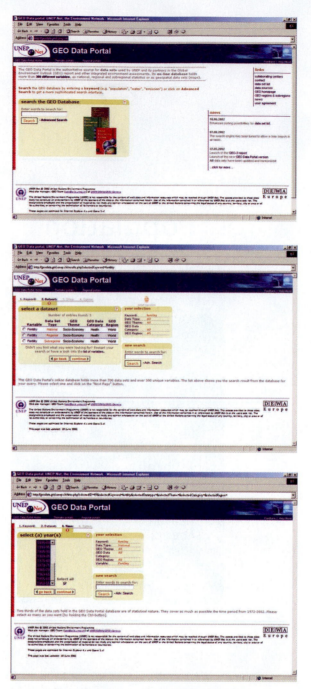

Screen shots from GEO Data Portal, UNEP 2002. Used with permission.

Step 2 Under "search the GEO Database," enter the word "fertility," and click "Search." You should now see a set of available database options relevant to "fertility," as in the middle screen shot above.

Step 3 Choose the top data option, fertility at the national level, by clicking on the radio button and then clicking "continue."

Step 4 From the year selections, check the box labeled "Select All" next to the list of available years, and then click "continue," as illustrated in the bottom screen shot above.

You should now be looking at a list of available output options for the data, as shown on the right. As a data "portal," the GEO Data Portal offers data to view in a map, chart, or table, as well as to download for use in statistical or mapping packages.

First, let's find out what type of data we've got by looking at the **metadata.** Metadata is the background information about a geographic data set. It includes facts, such as the source of the data, the scale at which it was collected, the year it was collected, the projection if there is one, and any other information that you need to know before you can interpret the meaning of the data.

Step 5 Under "Show Metadata," click "display as...Metadata."

Question 1: Read the "Abstract" and "Purpose" sections of the metadata. How is fertility rate defined for this data set?

Question 2: How was the data for fertility rate collected and measured?

Question 3: Why is fertility rate considered a more useful population indicator than birth rate?

Step 6 When you're finished browsing the metadata, click "go back" at the bottom of the page to return to the display options page.

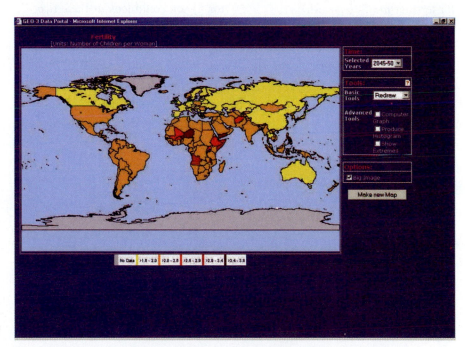

Screen shots from GEO Data Portal, UNEP 2002. Used with permission.

Step 7 Under "Draw Map," click on the image of the map. This will open up a separate window with a world map showing estimated fertility rate for the years 2045–50, as in the screen shot above.

Step 8 To get a good view of the data, resize the window of the data portal to fill your monitor screen. Then, under "Options," check "Big Image," and then "Make new Map." You should now have a map large enough to see some detail.

The fertility rate map shows a century of estimated data for each country. How are regional patterns of fertility estimated to change over this period of time?

Step 9 Explore the different estimates by selecting another time period from the "Selected Years" drop-down menu, keeping "Basic Tools" on "Redraw," and then clicking "Make new Map."

Question 4: Choose four different time periods from the drop-down menu and analyze what you see. What regional patterns do you find for fertility rate?

Question 5: Based on these patterns, which countries or regions might you predict to have a decreasing population density?

Hint: By selecting the "Identify" tool and then clicking the map, you can get data for the individual countries.

Step 10 Close the interactive map window. You should be back to the options screen as in the top image on page 93.

Step 11 Next, go back and explore the global data for Infant Mortality Rate. Click on "1. Keyword:" in the upper left section of the Web page. This should take you back to the view as in the topmost screen shot on page 92. In the box, type "infant mortality" and click "Search."

Step 12 From "select a dataset", choose "Infant Mortality Rate ~ National," click "continue," again choose all years of the data, and click "continue."

Step 13 Draw your map as in Step 7.

Question 6: Browse the estimated infant mortality data between 1950 and 2050. What regional patterns do you see?

Question 7: Reflect on what you have learned in class about infant mortality rate as a population indicator. If you could look at these two datasets, infant mortality and fertility rate, simultaneously, how would you expect them to correlate? In other words, for a country with a high fertility rate, would you expect infant mortality to be high or low? Explain your reasoning.

7.2
Population Indicators II: Exploring the Details

In exercise 7.1, you got your feet wet exploring population indicator data generalized at the global scale. How do these population dynamics shift at larger, regional scales?

In this exercise, we will zoom in on Europe, to an online interactive map called Descartes. Descartes is a powerful tool for exploring the correlation of geographic data at the country, city, and district level, through thematic maps and statistical charts of all kinds. (A newer version of the software, called CommonGIS, can be downloaded for educational purposes from www.commongis.de; for this exercise, we will use the online version.) The Descartes Web site is useful for exploring more specific aspects of geodemographic data at the local and regional levels.

Note: This Web site uses a particular type of network connection that may not be acceptable by all networks. If you are unable to work on this Web exercise from your computer, try accessing the site from your campus network or another type of network connection available to you.

Step 1   Launch your browser and navigate to the Descartes Web site at **http://allanon.gmd.de/ and/java/iris/**.

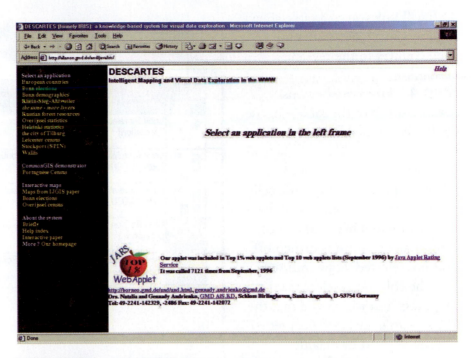

Screen shots of Descartes Web site, © 2002 Dr. Hans Voss, Spatial Decision Support Team. Used with permission.

Step 2   From the list of applications listed in yellow in the left frame, select the first option, "European countries," and choose "English" from the language option buttons. This will link you to a list of available options depicted to look like file folders, as shown in the bottom image.

Step 3   From the list of available data sets for Europe, choose "Relative indexes" and click "Show," as in the bottom image, above.

This will link you to a table showing population indicators for each country in the region.

Step 4   From the buttons across the top of the table, choose "Select column(s)."

This will take you to a window where you can select specific cells representing the columns of data to visualize in a map. Gray indicates that a column is turned off and not selected; aqua indicates that the column is selected for mapping. You will notice that the "European countries" data set is already selected by default.

Let's begin by comparing birth rate and population density across Europe. Before making your map, however, think about what you have learned about birth rate and population density from your textbook.

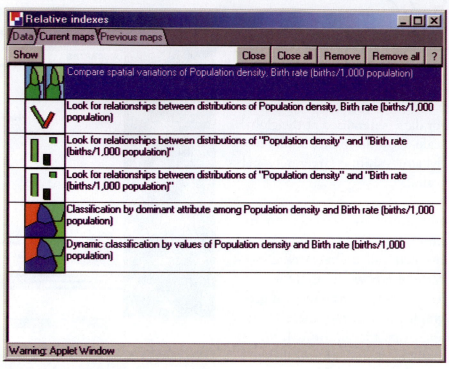

Screen shot from Descartes Web site, © 2002 Dr. Hans Voss, Spatial Decision Support Team. Used with permission.

Question 1: How are the two indicators birth rate and population density related? In other words, if population density is high, would you expect birth rate to be high or low?

Step 5   From the list of data, select "Population density" and scroll to the right to select "Birth rate (births/1,000 population)." Make sure the "Visualize" box is checked below the list of cells, and click "OK."

The result should be a list of different ways that this data can be visualized, called "Relative indexes."

Step 6   From the list of relative indexes, choose the second option, which will map the data using two diverging bars. Blue indicates that the option has been selected. Next, click "Show." Finally, a map!

Step 7   Adjust the size of the window so that your map fills the screen. In the frame to the right of the map, you should see a default list of tools available for data exploration. Change this frame from "Tools" to "Legend" by clicking on the Legend tab.

Step 8   Use the legend to interpret the two data sets shown in the map.

Question 2: What pattern do you see? Does this pattern match your expectations of the relationship between birth rate and population density?

Next, let's explore the **age distributions** of this region.

Step 9   Close the birth rate map, and then close the "Relative indexes" window, to reset the functions of the software. (You should be back to the view shown in the bottom screen shot on page 95.)

Step 10   Select "Relative indexes" again by clicking on "Show," to open the full table of relative population data.

Step 11 Examine the columns in the data table under "Part (%) in total population." This is the **age distribution** data for each country.

Question 3: Which country has the youngest population? Which has the oldest population?

Step 12 Next, click on "Select column(s)" again to build the layers of the map. This time, select the age distribution data sets "Part (%) in population" (select the subcategories of 0–14, 15–64, and 65 and over, but not the female and male categories). Also, for comparison, select the "National product per capita, $" and click "OK."

Step 13 In the "Relative indexes" table of map options that results, choose the first option ("Compare spatial variations...") and click "Show." This will result in a window showing four small maps of Europe in the center frame, as in the screen shot above, right.

Step 14 Before you start exploring the maps, make the display as large as possible on your monitor. You can increase the size of the maps by clicking and dragging the borders of the frame outward, giving more space to the center frame.

Step 15 Finally, click the "Legend" tab in the right frame to show the legend information instead of the toolbar.

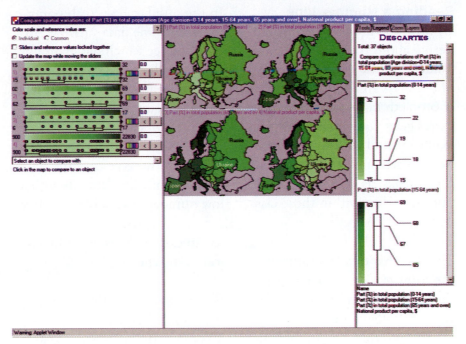

Screen shot from Descartes Web site © 2002 Dr. Hans Voss, Spatial Decision Support Team. Used with permission.

You should be looking at a map view as the one in the screen shot, above.

Step 16 Begin exploring the data shown in these maps.

The first three views of Europe show the age distribution data. The fourth map shows **GNPpc** for each country.

As you move your cursor across the maps, you will notice that if you position your cursor over one country in one map, that same country will be highlighted in each of the maps, to help you make quick visual comparisons of the data sets. Although not all of the countries are labeled in the map, you will notice that as you move your cursor across each country, the name of that country and its corresponding data categories are shown in the bottom corner of the right frame.

Question 4: Consider the relationship between the age distribution and the GNPpc. Are countries with younger populations showing a higher GNPpc, a lower GNPpc, or is there no pattern of correlation?

Question 5: Is there a pattern for countries with older populations? Explain your findings. Is this the pattern you expected to find, based on your knowledge of age distribution and wealth from the textbook?

Reading a Number Line

By now, you are probably wondering about the dots, bars, and "sliders" in the graph to the left of the map. In addition to the choropleth maps themselves, Descartes gives you many options for exploring and adjusting the classification scheme, class widths, and class breaks, as well as the distribution of data within those class widths.

To show that data distribution, the software gives you a graph that is particularly suited to choropleths called a **number line** (sometimes called a number-line plot). In this case, a number line is a way of plotting each data value along a single axis, so that the reader can see the distribution of data values within each of the data classes.

For example, imagine that you are looking at a choropleth map with this data legend:

▢	0–20
▢	21–40
▢	41–60
▢	61–80
▢	81–100

Although the legend appears symmetrical, the data set that it represents is not symmetrical at all but highly clustered. To show the reader the actual distribution of the data, the cartographer could include a number line:

In the number line, each point represents one data variable. To help you see which data points belong with each data category, the points are colored according to the classification category in which they belong.

Although this type of graph is effective only for data sets with a low number of variables (so they can all fit on the line), it makes for an extremely useful companion to the choropleth.

For example, sometimes the classification scheme of a map creates categories with wide gaps inside the class category, or the data set as a whole includes outliers, data values that are not clustered close to the rest of the values in the set. In these circumstances, the number will provide that information to the reader as a supplement to the map.

Question 6: Describe the distribution of data in the sample data set below.

Question 7: What do you think would be a better classification scheme for this data if it didn't have to be in equal intervals?

Reevaluating the Choropleth

Now you are ready to bring together all of your choropleth skills to reevaluate your findings about European geodemography.

Step 17 Backtrack through the windows in Descartes so that you are back at the file folder of data set options for European countries (see Step 2):

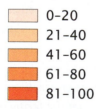

Step 18 As you have done previously, choose "Relative indexes" and click "Show" to return to the full table for this data set.

Step 19 Click "Select column(s)," choose just the data set "Population growth rate (%)," and click "OK."

Step 20 From the Relative indexes window, choose the third visualization option, "Dynamic classification by values..." and click "Show." Maximize the map window that results.

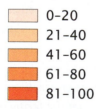

You should be looking at a map of a single data set, with a number line across the top of the window, as in the image on the right.

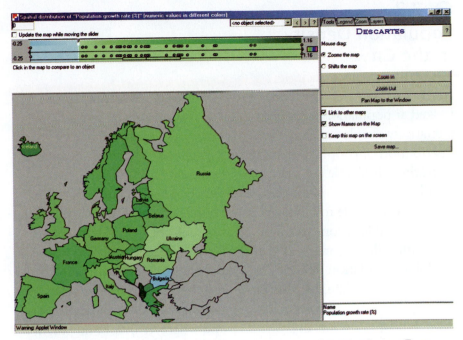

Screen shot from Descartes Web site, © 2002 Dr. Hans Voss, Spatial Decision Support Team. Used with permission.

Step 21   Click the Legend tab to display the legend in the right frame, and study the information that it provides about the map.

Question 8: Based on the information in the legend, how many data classes have been created for this data set?

Question 9: What is the classification scheme? (To review the different types of classification schemes, refer to Chapter 5.)

Question 10: Based on this classification scheme, what is the general distribution pattern of the population growth rate in Europe?

Step 22   To compare your impression of the data distribution to the data set itself, focus on the number line, which shows the data values plotted individually. Notice that the countries in the map are dynamically linked to the plotted data values, to help you make direct connections between map and graph.

Question 11: Which countries are the outliers in this data set?

Question 12: Are the outliers represented in the distribution shown in the map?

Step 23   Redraw this map using a **natural breaks classification scheme.** You can do this without going to the data table, simply by examining the clustering of the data on the number line. In natural breaks, data classes reflect breaks in the data set.

To change the class breaks of the data classes, click and drag on the slider to move it up or down the graph. To change the number of data classes, click on one of the buttons numbered "2" through "9" to the right of "N:" above the map.

Question 13: How many data classes did you choose?

Question 14: What are the class breaks of your new classes? Explain why you chose this new pattern.

Question 15: Describe the geographical distribution of the population growth rate in Europe as depicted by your new classification scheme. How does this pattern compare with the pattern you observed in Step 3?

Question 16: How useful would your new choropleth map be without the number line? Do you still need the number line for a useful interpretation of this map?

✺ 7.3
Population Density in the City

In the first two exercises, you looked at population density as a relatively simple concept: the distribution of people over space. Even though people are constantly moving around on the earth, it is possible to represent their general locations through population density maps. A map of global population density shows, generally, the distribution of people per country, even though people are moving between those countries.

The problem is that at larger scales (county, city, or neighborhood), mapping the locations of population densities becomes more slippery because the movement of people increasingly becomes a relevant factor to consider.

For example, we can map the population density of Michigan by county, as shown in the map below. The map shows general population density by residence in each county. Because people are continuously traveling between the counties, the map doesn't necessarily show where people actually are during the day. It shows the density if every person was at home, whatever and wherever that home might be. So, the map shows a kind of theoretical density of people that never technically exists.

The larger the scale of the map, the harder it is to find an accurate representation of population density because of human movement.

If the map is based on the U.S. Census, the data "population density" depicts the locations of individual or family dwellings. A census population density map is actually a **dwelling density map** of the city.

So how do we map population density at the city level? Does dwelling density portray an accurate picture of people in the city? Is there a better way to visualize urban population in a map?

Urban geographers have long considered the changing nature of the city depending on the time of day. The day city has different inhabitants, with different activities, than the night city, and the weekend city is different again. These changing dimensions of the city by day and night are central to interpreting urban social structure.

But can the actual shifting locations of individuals themselves be displayed and analyzed with a map?

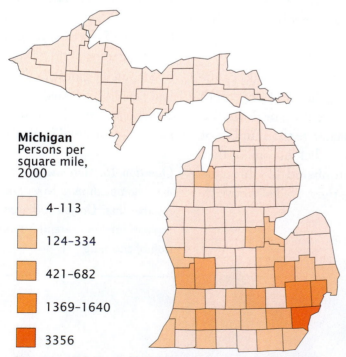

Michigan
Persons per square mile, 2000

- 4–113
- 124–334
- 421–682
- 1369–1640
- 3356

Data source: U.S. Census 2000. Base map data source: ESRI 1999.

This is a question that cartographer Janos Szegö has sought to answer. Szegö looked for a more complex picture of urban population dynamics, one which would go beyond dwelling density. He considered that the average person's life is divided into two broad categories of activity: work and home. How do those spaces compare, and how do they impact the urban infrastructure? To visualize the contrasting densities, he used a particular type of thematic map, the **isarithmic map.**

In this exercise, you will follow Szegö's isarithmic maps to discover a deeper picture of population density in and around the city of Kalamazoo, in southern Michigan. In the steps that follow, you will need a calculator for some simple arithmetic.

The Isarithmic Map

An **isarithmic map** depicts quantitative data as a smooth, continuous surface, using lines to connect numbers of equal value. The numbers of equal value might be point data, for example, surveyed elevation points. The numbers could also be data aggregated by enumeration unit, as in population numbers collected by census division.

One common type of isarithmic map in the United States is the topographical map series of the U.S. Geological Survey (USGS). A USGS "topo" is an isarithmic map because it uses contour lines

to show a continuous surface of elevation data. In chapter 2, you analyzed an early topographical map from the United Kingdom on pages 14–15, and from the United States on page 19. National weather maps in newspapers also commonly use isarithms to depict rainfall or temperature.

But isarithms can also be used to depict human geographical information. Because they show data as a continuous surface, isarithmic maps are useful for visualizing the volume of flow of people as they move around a city. In the next few maps, we will explore the extent to which isarithmic mapping can help us understand flows of people in Kalamazoo.

Question 1: To get oriented, study the locator map for Kalamazoo, below. Examine the street pattern and landmark information provided in the map. Where would you say downtown is? Where do you think the industrial areas are? What in the map gives you this impression?

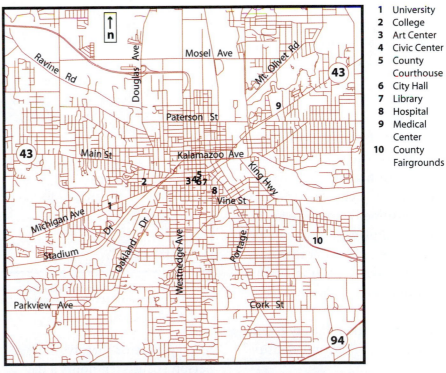

1 University
2 College
3 Art Center
4 Civic Center
5 County Courthouse
6 City Hall
7 Library
8 Hospital
9 Medical Center
10 County Fairgrounds

Base map data source: Michigan Geographic Data Library (www.michigan.gov/cgi).

Question 2: Now that you are somewhat oriented in the city, study the isarithmic maps for the city on the left. In which parts of town is dwelling density highest? Where is it lowest? Where is worker density highest? Lowest? (You can use the landmarks in the locator map to help explain your answer.)

Question 3: How does worker density compare to dweller density? Are there places where both maps show high densities?

Question 4: How does this comparison of dweller and worker densities compare to your impression from the locator map? How has your impression of the city changed?

Question 5: These isarithmic maps show the two major activity fields in a person's day: activity at work and activity at home. What other fields of activity not included here comprise daily urban life?

Once we have information about locations of workers and dwellers, we can use the map to extract information about the flow of people in the city. For example, if we assume that people are at their workplaces during the day and at their homes at night, we can make the further assumption that dweller density resembles the city's "nighttime density," and worker density resembles the city's "daytime density."

Dweller density, Kalamazoo 1998

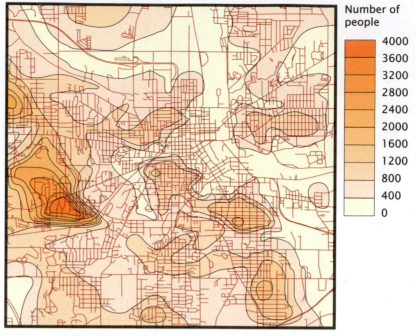

Number of people

| 4000 |
| 3600 |
| 3200 |
| 2800 |
| 2400 |
| 2000 |
| 1600 |
| 1200 |
| 800 |
| 400 |
| 0 |

Data source: Courtesy of Kalamazoo Area Transportation Study (KATS).
Base map data source: Michigan Geographic Data Library (www.michigan.gov/cgi).

Worker density, Kalamazoo 1998

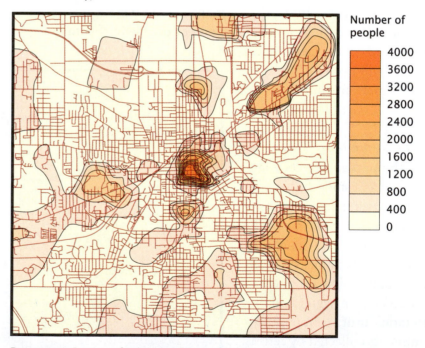

Number of people

| 4000 |
| 3600 |
| 3200 |
| 2800 |
| 2400 |
| 2000 |
| 1600 |
| 1200 |
| 800 |
| 400 |
| 0 |

Data source: Courtesy of Kalamazoo Area Transportation Study (KATS).
Base map data source: Michigan Geographic Data Library (www.michigan.gov/cgi).

Szegö estimated that, typically, a person spends ten hours a day at work (8 a.m. to 6 p.m.) and fourteen hours a day at home (6 p.m. to 8 a.m.). This information can be used to convert the isarithms in the first map from the "number of people" to the number of "personhours" spent in each space.

Nighttime Personhours, Kalamazoo 1998

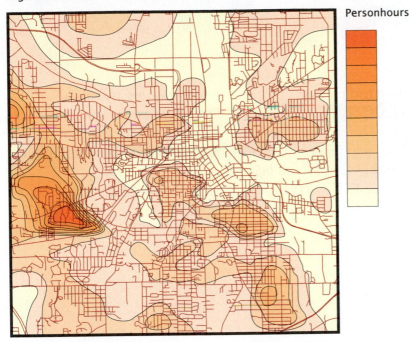

Personhours

Data source: Courtesy of Kalamazoo Area Transportation Study (KATS).
Base map data source: Michigan Geographic Data Library (www.michigan.gov/cgi).

Daytime Personhours, Kalamazoo 1998

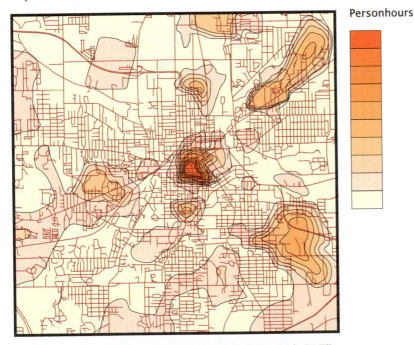

Personhours

Data source: Courtesy of Kalamazoo Area Transportation Study (KATS).
Base map data source: Michigan Geographic Data Library (www.michigan.gov/cgi).

For the dweller density map, the formula for personhours is: **Number of people × 10.**

For the worker density map, the formula for personhours is: **Number of workers × 14.**

Question 6: Convert the population density isarithms in the Dweller Density map and the Worker Density map to personhour density isarithms. The isarithms themselves will not change; all you have to do is create new labels for each legend category on the right side of the legend boxes (use photocopies if you are handing in this exercise).

Question 7: Compare your two new personhours maps. How do they compare?

Question 8: Why do you think an urban geographer or planner would be interested in day and night maps of personhours?

Question 9: What are the limitations of using the data to make assumptions about time of day and personhours? (Hint: Think about your answers to questions 4 and 5.)

To Szegö, understanding where people are in a city is the most fundamental awareness that one can have about how a city works:

What are the networks that create the relationships between the different components of a town? The answer is simply one word: people. It is people and the networks created by their lifelines, spun like webs around buildings, between buildings and squares, from small rooms to lofty galleries, private gardens and parks, that create the invisible structure of a living town. [Szegö, 1994: 218]

Sources and Suggested Readings

Classification Schemes

Monmonier, Mark. *Mapping It Out: Expository Cartography for the Humanities and Social Sciences.* Chicago: University of Chicago Press, 1993.

Population Indicators

Andrienko, G. L., and N. V. Andrienko. "Interactive Maps for Visual Data Exploration." *International Journal of Geographical Information Science* 13 No. 4 (June 1999): 355–374.

Population by Isarithm

Dent, Borden D. *Cartography: Thematic Map Design.* Fifth ed. N.Y.: WCB McGraw-Hill, 1999.

Szegö, Janos. *Mapping Hidden Dimensions of the Urban Scene.* Stockholm: The Swedish Council for Building Research, 1994.

8

Agriculture

Vocabulary applied in this chapter
agricultural landscape
agricultural region
agricultural diffusion
agricultural differentiation
environmental perception
Public Land Survey System
  (PLSS)

New vocabulary
satellite imagery
resolution
AVHRR
Landsat TM
SPOT
wavelength
electromagnetic spectrum
infrared

✳ 8.1
Agricultural Landscapes from Satellite Imagery

One way to monitor changes in **agricultural land use** and **landscape** pattern is through interpretation of **satellite imagery**. In this exercise, we will look at how satellite images can be used to explore different agricultural types and monitor or predict ecological change.

Types of Imagery

Unlike an air photo, which is a photograph taken with a camera and typically collected from an airplane, a satellite image is a digital image of some part of the earth's surface, collected by a sensor on a satellite. Satellite images differ by resolution and wavelength.

Each type of satellite image has its own **resolution.** The resolution of an image is comparable to the scale of a map: it defines the measure of detail of geographical information in the image.

For example, **Advanced Very High Resolution Radiometer (AVHRR)** images have a resolution of about a kilometer. This means that an object that measures one kilometer or more on earth will be visible in the image.

Landsat Thematic Mapper (Landsat TM) images have a resolution of 30 meters, meaning the images are made up of rectangles (or "pixels") 30 meters × 30 meters. This resolution comes out to a little less than a quarter of an acre. **SPOT** images, on the other hand, have a higher resolution of about 10 meters.

Like scale, resolution limits what kind of information can be shown. A farmer with one-acre fields, for example, would not find Landsat TM imagery useful because each field would be reduced to four pixels in the image, not enough to contain usable information. On the other hand, the same farmer might find Landsat TM imagery useful for understanding the agricultural patterns of the surrounding watershed or ecosystem.

Wavelength also influences the type of information that can be depicted in an image. Each satellite image is a picture of radiation from a particular part of the **electromagnetic spectrum.** Remote sensing scientists decide which part of the spectrum will be used in order to highlight certain features of the earth in the image.

For example, to map the spatial pattern of agriculture, geographers use satellite imagery from the **infrared** wavelength. Infrared allows image analysts to interpret the health and development of plants by measuring their chlorophyll reflectance levels.

Agricultural Systems in Border Landscapes

Study the two Landsat TM infrared images on the facing page. Both images depict parts of North Korea during August 2001.

In these images, the bright, reddish orange indicates paddy rice agriculture, darker orange indicates corn, yellow and green indicate other vegetation, and grey and white indicate urban areas. In the top image, you can see the orange of corn during harvest season in the vicinity of Pyongyang. In the bottom image, from coastal North Korea, paddy rice farming dominates the image.

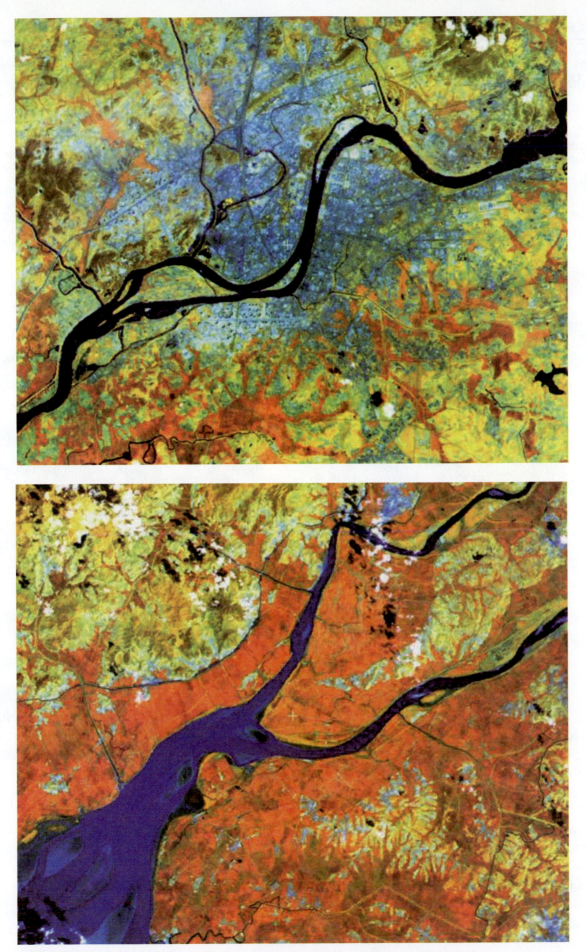

Landsat TM images, 2001 © USDA / Foreign Agricultural Service. Used with permission of USDA / FAS.

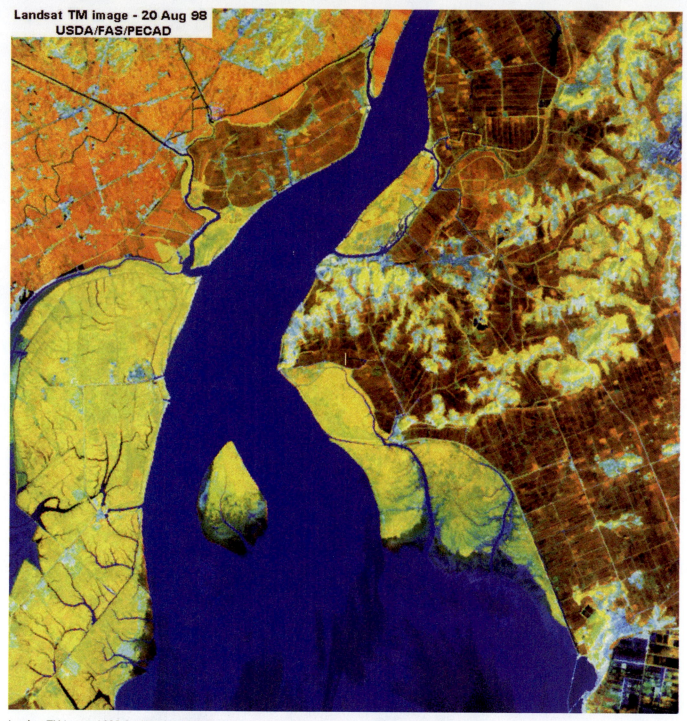

Landsat TM image - 20 Aug 98
USDA/FAS/PECAD

Landsat TM image, 1998 © USDA / Foreign Agricultural Service. Used with permission of USDA / FAS.

Satellite images can also be used to compare differences in agricultural practices between cultures, as in the next two Landsat TM infrared images. The above image shows the border of North Korea and China in August 1998.

Question 1: Reflect on the different types of agricultural systems that you have learned are characteristic of agricultural regions. What difference in agricultural systems between North Korea and China is depicted in the image?

NORTH and SOUTH KOREA

NORTH KOREA

SOUTH KOREA

LANDSAT TM 91/10/22
USDA/FAS

Landsat TM image, 1991 © USDA / Foreign Agricultural Service. Used with permission of USDA / FAS.

The image on this page shows the border between North and South Korea in October 1991. The coloring of this Landsat TM image is slightly different: unharvested paddy rice is depicted in white.

Question 2: What differences in agricultural systems between North and South Korea can you see in this image?

Monitoring Agricultural Conditions and Change

In 1997, North Korea suffered a devastating drought, the effects of which continue to influence North Koreans today.

The Foreign Agricultural Service (FAS, a division of the USDA) has been monitoring the agricultural conditions in North Korea in order to monitor, predict, and preempt such tragic consequences in the future. To make those kinds of predictions, the FAS relies on SPOT imagery.

Compare the two SPOT infrared images of Huanghae Province, below. Both images are from the month of August, in the years 1998 and 1997, respectively.

In these images, vegetation appears as red. The higher the levels of chlorophyll in the vegetation, the brighter the red of the image.

Question 3: What evidence can you find for the previous year's drought, as shown in the second image?

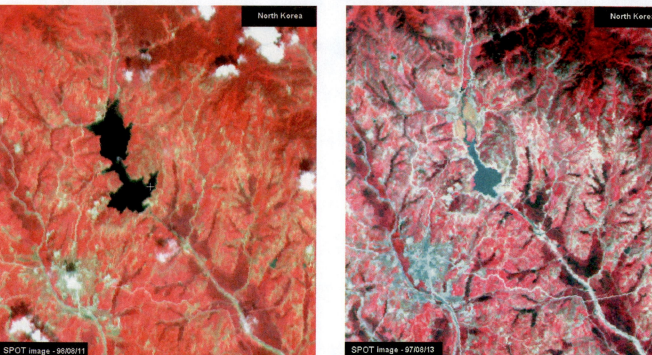

SPOT images, 1998, 1997 © USDA / Foreign Agricultural Service. Used with permission of USDA / FAS.

In this final infrared image from SPOT, the border between China and North Korea is shown for July 1997.

Question 4: What differences for the two countries can you see?

Question 5: How does a political border influence agricultural conditions during a drought?

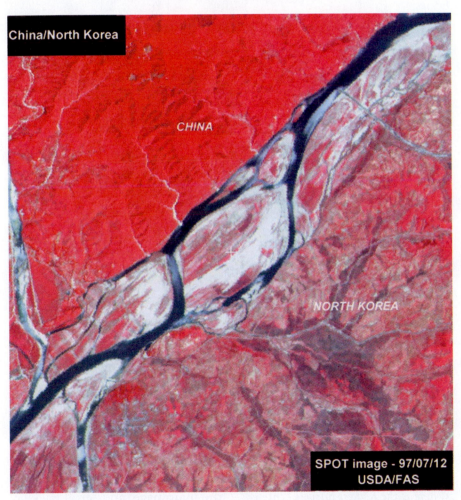

China/North Korea

CHINA

NORTH KOREA

SPOT image - 97/07/12
USDA/FAS

SPOT image, 1997 © USDA / Foreign Agricultural Service. Used with permission of USDA / FAS.

🔍 8.2
Diffusion and Differentiation in Agricultural Regions

How does the geography of a crop change over time? How do you track **agricultural diffusion** cartographically?

This exercise utilizes the online mapping of the National Agricultural Statistics Service (also part of the USDA). The NASS Web site allows you to query historical data for major crops in the United States, both in terms of total acreage harvested and product yield per acre. The online maps at this site allow us to go beyond generalizations about major agricultural regions and deeper into the details about processes within and between those regions.

Step 1   Launch your browser and go to the National Agricultural Statistics Service at **www.usda.gov/nass/aggraphs/ cropmap.htm.** You should find yourself at the "Crop Acreage and Yield" Web site shown in the screen shot, top right.

Step 2   Scroll down to the map of "Sunflower Harvested" and click the map. This will take you to a choropleth map for the most recent harvest, as in the bottom screen shot on this page.

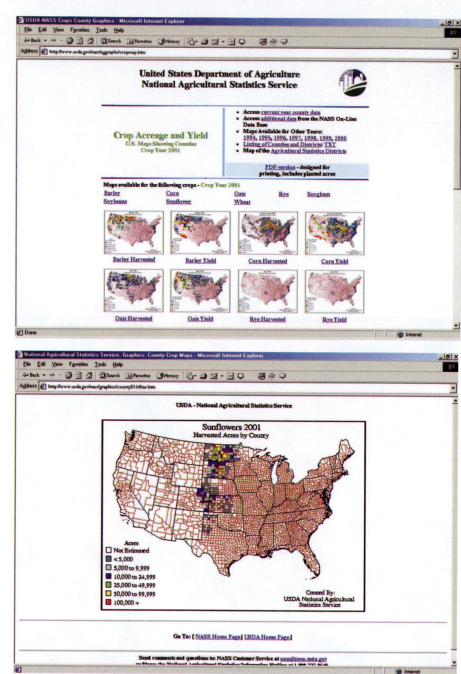

Screen shots from National Agricultural Statistics Service (NASS), U.S. Department of Agriculture Web site, 2002. Used with permission.

Question 1: Describe the regional distribution of the sunflower harvest. Where is the highest harvest centered? What are the boundaries of the sunflower-growing region?

Step 3   Hit the "Back" button on your browser to go back to the main page, and scroll back to the top of the page to the "Maps Available for Other Years" section.

Is there evidence of agricultural diffusion in the sunflower crop? Explore the past geographies of the sunflower harvest by clicking on the years for 1996 to the present.

Question 2: Compare the change in distribution across the maps. How has the geography of harvested acres of sunflowers changed since 1996?

Question 3: What factors do you think may have contributed to the change in the sunflower-farming region over time?

The NASS data is also useful for tracking **agricultural differentiation,** specifically, regional variation in agricultural productivity and price. Durum wheat, an essential ingredient to high-quality pasta, is an interesting crop to track graphically.

Step 4   Go to the data for 1999 and scroll down to the maps for "Durum Wheat Harvested" and "Durum Wheat Yield."

Question 4: What, in general, is the difference between the acres planted in durum wheat in North Dakota versus Southern California? Which region has the highest number of acres planted in this crop?

Question 5: Compare your findings for harvest data to the information in the yield map. Which region has the highest yield of durum wheat?

Question 6: What do you think might be some factors contributing to this regional disparity between yield and acres harvested for this crop? Has this disparity affected the market price? (Hint: Market prices are available through the "Additional data of the NASS" link at this site.)

⟳ 8.3
Environmental Perception, Agriculturists, and the Map

Although maps have the power to alter our **perception** of agricultural regions and landscapes, that power is not always successfully interpreted and used by the map reader, or by society. A good example of this comes from a map by John Wesley Powell, former director of the United States Geological Survey (USGS).

In 1878, Powell presented the findings of his extensive research on the Great Basin in his work *Report on the Lands of the Arid Region of the United States.* Through written and cartographic narrative, Powell documented the arid nature of the West and warned against the popular perception that the Western environment could support large-scale agricultural expansion. Powell reported that expansion could proceed, but in order to preserve the agricultural health of the region, the federal government would have to proceed with specific, water-conscious measures.

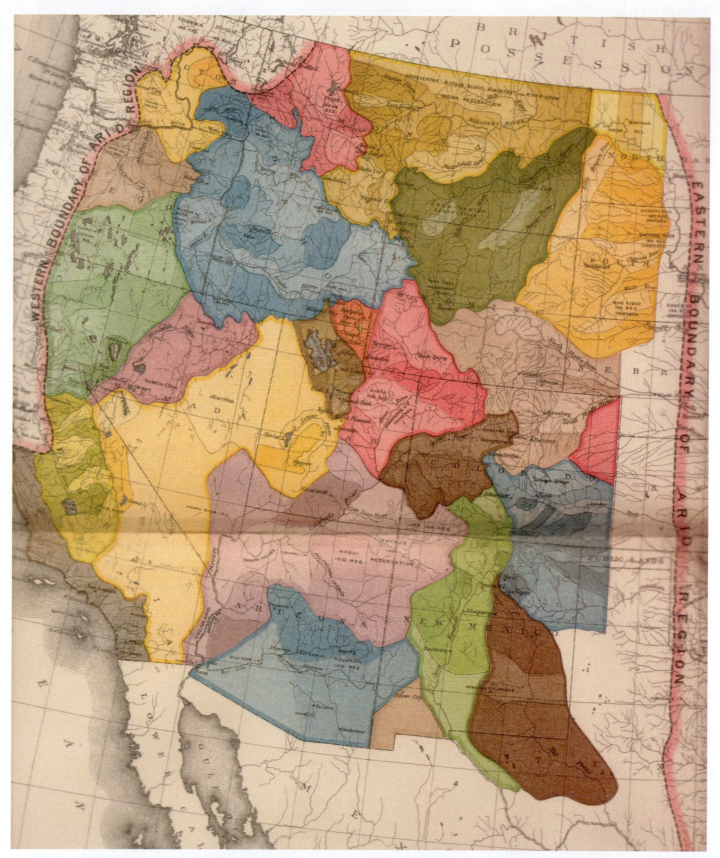

"Arid Region of the United States Showing Drainage Districts" by John Wesley Powell. *USGS Eleventh Annual Report Part II—Irrigation*, Plate LXIX. Washington: GPO, 1891.

In particular, Powell hoped to convince the Department of the Interior that the settlement of the West would be sustainable only if political jurisdictions within the states were based on watershed boundaries, rather than the arbitrary gridded lines of the **Public Land Survey System** (PLSS). Powell perceived that watersheds would best be preserved if neighbors with common interests cared for the watersheds at the community level. With communities built on the arbitrary grid of the PLSS, those watersheds would be divided across many counties.

Two years later, in a follow-up report, Powell illustrated his ideas in a detailed map, titled "Arid Region of the United States Showing Drainage Districts," shown on page 114. In the map, he graphically demonstrated that political districts created by watershed boundaries would organize the region into manageable units.

In this case, the map failed to alter the environmental perception of agriculturists. Powell's watershed district proposal was ignored by the Department of Interior, and Powell resigned from his position as Director of the USGS two years after the publication of 1891 report. Perception of the West as a region of agricultural expansion and opportunity continued to pervade the American imagination, and counties and agricultural property boundaries were based on the grid of the PLSS as planned.

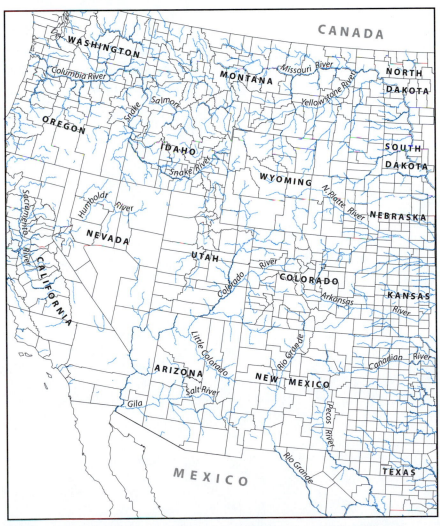

Base map sources: AWIPS Map Database, National Weather Service, and ESRI 1999.

Question 1: Compare Powell's map to the present-day county boundaries and rivers in the map on the left. Are there any counties that follow parts of the watershed? If so, in which state?

Question 2: Imagine that Powell's ideas had been implemented and Western jurisdiction was based on the watershed unit rather than the county unit. Which state do you think would be most different today and why?

Question 3: What do you think the overall agricultural impact would have been if Powell's jurisdictions had been implemented?

Question 4: Do you think there would be drawbacks to Powell's plan? Explain your reasoning why or why not.

Question 5: What factors determine whether a map successfully alters public consciousness about a geographical theme?

Sources and Suggested Readings

Satellite Imagery

Lillesand, Thomas M., and Ralph W. Kiefer. *Remote Sensing and Image Interpretation*. 3rd ed. N.Y.: Wiley, 1994.

Agricultural Diffusion

Pillsbury, Richard. *Atlas of American Agriculture: the American Cornucopia*. N. Y.: Macmillan, 1996.

John Wesley Powell

Powell, John Wesley. "Arid Region of the United States Showing Drainage Districts." *United States Geological Survey, Eleventh Annual Report, 1889–90: Part II, Irrigation Survey*. Washington, D.C.: Government Printing Office, 1891.

Powell, John Wesley. *Report on the Lands of the Arid Region*. Washington, D.C.: Government Printing Office, 1878.

Stegner, Wallace. *Beyond the Hundredth Meridian: John Wesley Powell and the Second Opening of the West*. N.Y.: Penguin, 1992.

9

Industry

Vocabulary applied in this chapter
industrial regions:
   formal, functional,
   vernacular
industrial ecology
industrial landscape
industry types:
   primary
   secondary
   tertiary
   quartenary
   quinary

New vocabulary
Geographic Information
   System (GIS)
Web GIS

9.1
Exploring Contemporary Industrial Regions

This exercise explores the geography of a quaternary **industrial region.** What are its boundaries, and which criteria should we use to describe its boundaries? Does this region have a center or node? To understand this region, we will use the capabilities of an online GIS, SanGIS.

Introduction to GIS
When mapping combines the analytical capabilities of the computer with the visualization benefits of the map, this is known as a **Geographic Information System** or **GIS.** A GIS consists of a computer, the software tools for performing GIS analysis, digital databases, and database analysis. Though people often print paper maps in order to display the results of their analysis, the greater strength of a GIS lies in the computer system itself, where the user can graphically examine different views of a spatial dataset as a means of developing a deeper knowledge of that data than afforded by nongraphic devices, such as tables or descriptive statistics.

Geographers use GIS for any kind of work in which the primary goal is to store, analyze, and graphically view spatial databases. A GIS is more useful than a paper map when you have data associated with each of the geographical features of a particular

place and you wish to perform statistical analyses of this feature data. The applications for GIS are wide-ranging and are used throughout physical and human geographical research, limited only by the availability (and quality) of data for a particular place.

When GIS is combined with the interactive capabilities of the Web, it is known as **Web GIS.** Web GIS has grown increasingly popular in recent years as an effective means of providing geographical data in an accessible format. Whereas a conventional GIS demands that a user have special software in order to access its geographical databases, software that typically requires special training for use, a Web GIS by contrast can be accessed with

nothing more complicated than a web browser. For this reason, it is ideally suited to the dissemination of data for geographical query to a wide range of users.

The City of San Diego, in partnership with the San Diego Geographic Information Service (SanGIS), runs an interactive Web GIS called the Bandwidth Bay Fiber Network Map intended for use by quaternary industries thinking of locating to San Diego. The network map is designed to demonstrate that San Diego has the communications, housing, and entertainment infrastructure necessary to support high-tech companies and their employees.

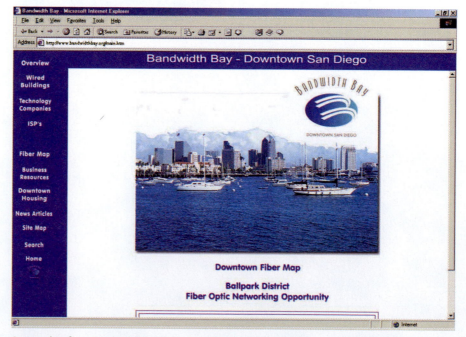

Screen shot from Bandwidth Bay Web site, 2002. Used with permission of SanGIS.

Step 1 Launch your browser and go to the Bandwidth Bay Web site at **www.bandwidthbay. org/main.htm.**

You will see the Bandwidth Bay welcome page as shown in the image on page 118.

Step 2 In the main frame, click on "Downtown Fiber Map." This will take you to a frame introducing and describing Bandwidth Bay. Scroll down and watch the video clip "Paradise in Progress" at this frame (this video has audio narration).

Next, scroll back to the top of the frame and click on "Downtown San Diego, Interactive Fiber Map." This will link you to a frame introducing you to the goals and capabilities of the Bandwidth Bay Fiber Network Mapping system.

Question 1: Based on this introductory information, describe the trait or traits that you think define the region "Bandwidth Bay."

Question 2: Is Bandwidth Bay a formal, functional, or vernacular region? Why?

Step 3 Scroll down to the bottom of the frame and click "To the Bandwidth Bay Fiber Network Map." This last link should take you to a frame with a Web GIS of the City of San Diego, as shown in the image above, right.

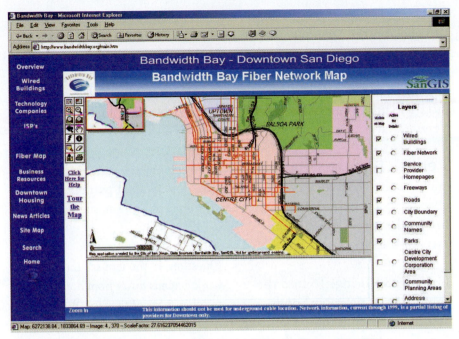

Screen shot from Bandwidth Bay Web site, 2002. Used with permission of SanGIS.

Step 4 Take a few moments to get oriented with this interactive map. In the upper left, a locator map shows you what part of San Diego you are currently viewing.

To the left of the locator map, a toolbar provides zoom, pan, query, and search tools. The selected tool is outlined in red and described in the bottom left corner of the map. The view should be set to its default tool setting with the "Zoom In" tool selected in red and indicated at the bottom left.

Directly below the locator map, a scale bar indicates distance at this particular view of the city.

On the right side of the map, the data layers available for this view are listed. Some data layers are available at only certain scales, so this list will change when you zoom in and out of the map.

Step 5 In the toolbar, click the "Legend/LayerList" tool (the top left icon on the toolbar).

This tool converts the layer list to a legend explaining the colors and symbols in the map. You can see that the fiber network is distinguished from the roads by a red network of lines, and "wired buildings" are depicted with a red mesh pattern.

Step 6 To get an overview of the fiber network itself, toggle back to the layer list with the "Legend/LayerList" tool. In the list of layers, click the radio button next to "Fiber Network" to make it "Active for Details." This makes the fiber network the active layer out of all the layers in the map.

Step 7   Return to the toolbar, and click on the "Zoom to Active Layer" tool (the third icon from top, on the right). In a few moments, the map will redraw to show the full extent of the Fiber Network layer.

Step 8   Take a moment to examine the geography of the fiber network, both downtown and in the extensions to other parts of the city.

Tips for exploring: To find the names of city neighborhoods, first make "Community Planning Areas" the active layer in the layer list. Return to the toolbar and click on the "Identify" tool. Then click your cursor over the neighborhood you would like to identify. The name of the neighborhood will appear in the query box below the map, as shown in the screen shot below.

Question 3: What part of the built environment seems to shape this network?

Question 4: You can see that the network extends away from downtown to other parts of the city. Which San Diego neighborhoods are served by the fiber network?

Step 9   Compare the extent of the fiber network to the locations of wired buildings. As you did for the fiber network previously, make "Wired Buildings" the "Active for Details" layer, and click the "Zoom to Active Layer" tool in the toolbar.

Step 10   Compare the extent of each layer by toggling back and forth using the "Back to Last Extent" tool in the toolbar. You can also use the locator map to help you compare visually.

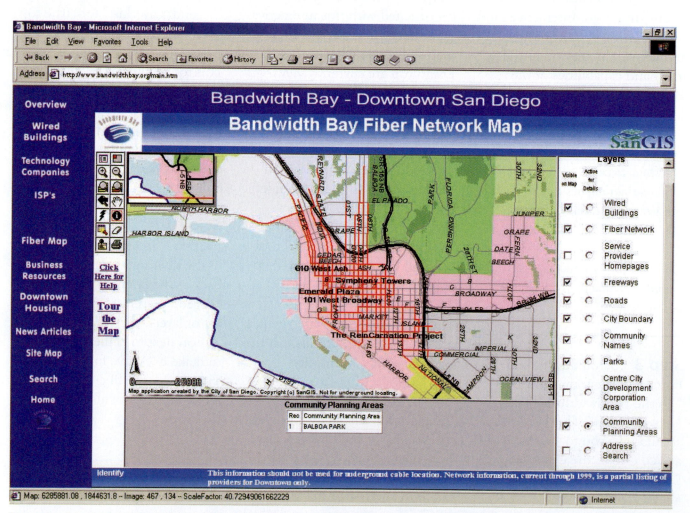

Screen shot from Bandwidth Bay Web site, 2002. Used with permission of SanGIS.

Question 5: How much of the fiber network is used by wired buildings?

Question 6: From your observations of the fiber network and wired buildings, and any other traits defining the region from your answer to Question 1 above, what do you think the boundaries of this region are?

Now that you have defined the region's outer boundary, let's find out if the region has a center or node.

Step 11   Toggle back to a view of the wired buildings layer if you are not presently viewing this map.

Step 12   Activate the "Aerial Orthophoto" layer, and scroll down to the bottom of the layers list and click "Refresh Map." Although there is no data associated with the orthophoto layer, an orthophoto helps visualize the cultural landscape.

Step 13   The Wired Buildings layer includes rental information about each building. Make the Wired Buildings layer "Active for Details," click the Identify tool, and click your cursor on a wired building. In the query box, you will see the street address, building size, building age, and vacancy rate of the building you queried.

Question 7: Which blocks or streets have the lowest vacancy rates? The highest vacancy rates?

Question 8: Based on this information, which area of downtown seems to have the most competition for office space?

In addition to building information, SanGIS provides data about which Internet Service Providers (ISPs) serve each segment of the network.

Step 14   To look at the information, make the Fiber Network the active layer, keep the Identify tool the active tool, and click your cursor on a segment of the fiber optic network shown in red.

Question 9: Which streets or blocks indicate the most competition between ISPs? Which indicate the least competition?

Remember that the fiber network extends out much further than the downtown area.

Step 15   To continue comparing ISPs, click on the "Zoom to Active Layer" tool to display the full extent of the network. Next, continue to query the network segments using the Identify tool. Toggle to the legend if you want to find the names of different regions of the city.

Question 10: What generalizations can you make about ISP service outside of Centre City?

Question 11: Based on your observations so far, compare the geographies of wired buildings and fiber networks. Do the two geographies correspond?

Question 12: Does a center or node for this industrial region emerge? Is there a periphery?

Question 13: Based on your reading of the Fiber Network Map, is there such a place as Bandwidth Bay? Explain your answer.

⊕ 9.2
Industrial Ecology on the Web

As you saw in the previous exercise, Web GIS is useful for revealing industrial regions. It can also be useful for revealing **industrial ecology.** The Environmental Protection Agency (EPA) uses Web GIS as one of several modes for disseminating environmental information to the general public. Its interactive Web site, "Window to my Environment," was created as a resource for citizens to make geographical comparisons between the physical context of their communities and the EPA-regulated industries located there.

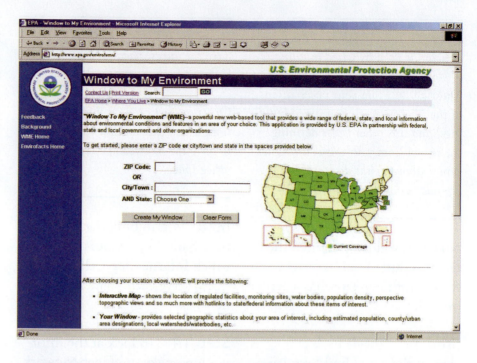

Step 1   Launch the "Window to My Environment" (WME) interactive map from the EPA Web site at **www.epa.gov/enviro/ wme/.**

You will see a map of the United States, with the current database coverage indicated on the map, as in the image above, right. This is the geographic gateway to using the EPA's interactive data browser.

Step 2   In the "Zip Code" box, type the zip code of your hometown, and then click on "Create My Window." If you do not live in a state with coverage, search for a town in a state that is covered. This will return a zoomed map of this area code, with streets, highways, and water features indicated, as in the bottom image on this page.

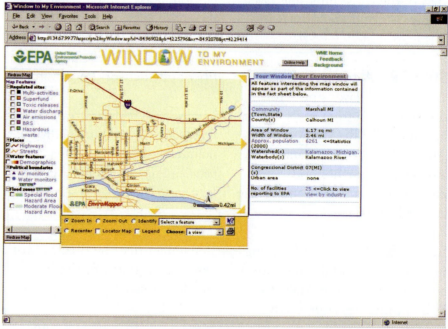

Screen shots courtesy of the U.S. EPA Office of Environmental Information. Used with permission.

To the left of the map, a column lists the map layers that can be turned on and off in the region you have chosen. Bold headings (Regulated sites, Places, Water features, Political boundaries, and Flood zones) indicate that an additional list of mappable features exists for that category. To view additional data layers, click on the "plus" signs to the left of the headings.

Step 3   Experiment with the data layers by exploring the layers listed under Regulated sites. Click on the boxes next to Superfund, Toxic releases, Water dischargers, Air emissions, and Hazardous waste, and then click on the "Redraw Map" button at the top or bottom of the column.

Question 1: Make a list showing how many of each EPA site type exist in your zip code area:

Type
Number

Superfund clean-up sites
Toxic releases
Water dischargers
Air emissions
Hazardous waste

Question 2: Look at the distribution of the EPA regulated sites. How are they clustered? Using your knowledge of your hometown, describe the distribution of these sites with respect to the physical and built landscape of your town or city. Use the Zoom In, Zoom Out, and Identify tools in the orange toolbox below the map viewer to assist with your interpretation.

Step 4   You can identify any of the sites by name at this site. Click on the "Identify" button from the tools in the yellow field below the map, and in the "Select a Feature" drop-down menu, choose the category of the feature you would like to identify. Next, click on that feature symbol in the map. The WME will return the name or names associated with that symbol, with a link to more information below the map.

The map provides a good picture of the types of EPA sites — but how does this relate to the major types of industrial activity? Recall from your textbook that there are five types of industry: primary, secondary, tertiary, quaternary, and quinary.

Step 5   Choose one of the EPA site types from your list in Question 1, above, turn off all the other layers, and click "Redraw" so only this site type is displayed.

Step 6   Using the "Identify" tool, browse the EPA information associated with each industry, noting particularly the type of industry (primary, secondary, etc.).

Question 3: What do you find for number of industry types? For example, if you chose Air Emissions, how many of those air emissions sites are primary industries? How many are secondary, tertiary, quaternary, or quinary industries?

Question 4: Examine your results from Question 3 above. Do these results change your perception of the environmental impact of industries in your town? If so, how?

Another layer of interpretation can be added by exploring the relationship between the EPA sites and population density.

Step 7   Click on the Demographics layer and redraw the map to view the population per square mile for this region.

Question 5: How are the EPA sites clustered with respect to population?

Question 6: Are there certain types of emitters that exist apart from the populated areas?

Question 7: Do sites exist inside the boundaries of the densely populated areas of town?

Notice to the right of the map, "Your Window" provides an overview of facts about your community. If you click on the other tab, "Your Environment," WME provides a list of specific questions concerning environmental data for your hometown, with links to assist you in finding answers.

These questions link to a wealth of environmental databases, fact sheets, and public and private agencies involved in the environmental health of this region. You can use them to locate environmental data as well as get involved in finding environmental solutions for your neighborhood. For instance, you can get a preliminary view of the types and amounts of chemicals released in your neighborhood.

Step 8   Under "Cross Media," click "What chemicals are released?"

Step 9   On the next Web page displayed, click "TRI Explorer." Note that, in the table that results, you can sort the chemical names alphabetically and the chemicals by quantity by clicking the arrow buttons at the top of each column. (Your browser may not automatically display the report in this step; if this is the case, click "Generate Report" to display a table of chemical release data.)

Question 8: What are the top three chemicals released on-site in your town? Off-site?

9.3
Exploring the Industrial Locations of the Past

One way to calculate fire insurance for urban areas is through extensive mapping of the building materials and urban context of every building. From the 1920s to the 1960s, fire insurance mapping in the United States was dominated by the Sanborn Map Company. Sanborn maps can be found for every city in the country.

Once used to calculate fire risk and changes in fire risk over time, Sanborn maps today are a gold mine of information about changes in the urban landscape. Highly detailed and large scale, the maps are sources for the names of building owners and the locations of businesses and residences over time. They also document the vanished, ancient, urban landscape, such as the paths of trolley and rail beds, the locations of stables, mill races, and even outdoor toilets.

Today these maps can be found in the archives of university libraries, city archives, and research libraries. At the University of Virginia, the Sanborn maps of Charlottesville, Virginia, are freely available online through the resources of the Geospatial & Statistical Data Center. This exercise utilizes the Charlottesville Sanborns as tools for interpreting the historical **industrial landscape.**

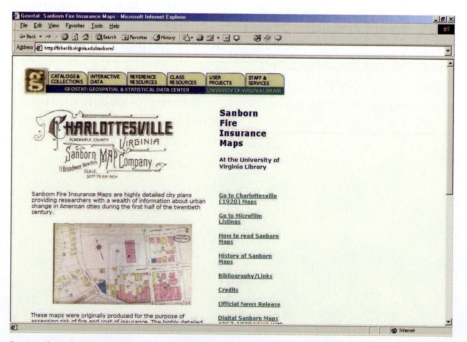

Screen shot, © Geospatial and Statistical Data Center, University of Virginia Library Web site (http://fisher.lib.virginia.edu), 2002. Used with permission.

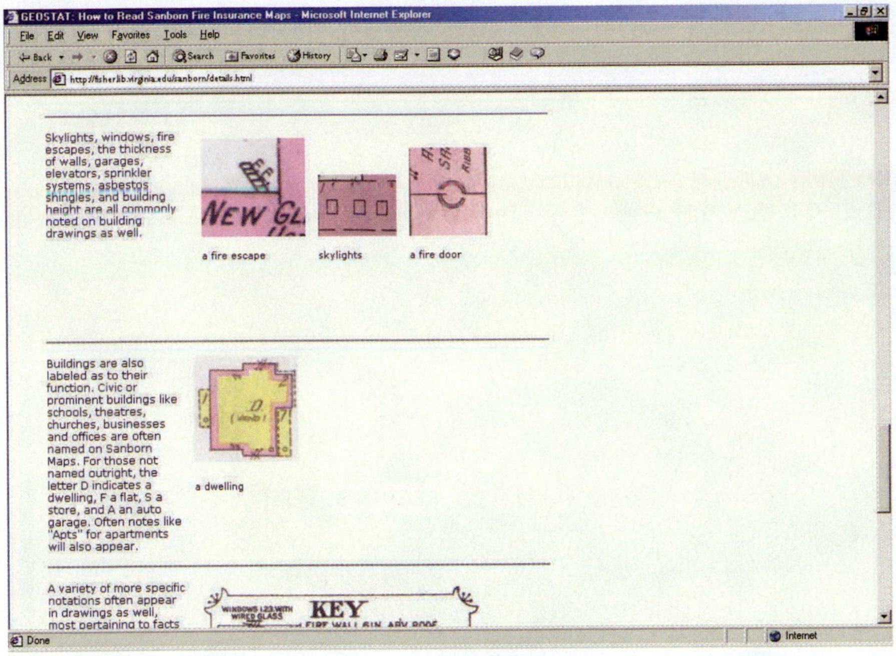

Screen shot, © Geospatial and Statistical Data Center, University of Virginia Library Web site (http://fisher.lib.virginia.edu), 2002. Used with permission.

Step 1   Launch your browser and go to the Sanborn Fire Insurance Maps page at the University of Virginia Web site at **http://fisher.lib.virginia.edu/sanborn.**

Step 2   From the right-hand column, click on "How to read Sanborn Maps." This link takes you to a brief, visual guide to interpreting the symbols and color coding of the Sanborn Maps, as in the screen shot above. Review the guide and study the explanation of symbols.

Question 1: Why did the Sanborn company map only parts of the city, rather than the city as a whole?

Question 2: How do you think this will limit interpretation of the cultural geography of industrial location in Charlottesville?

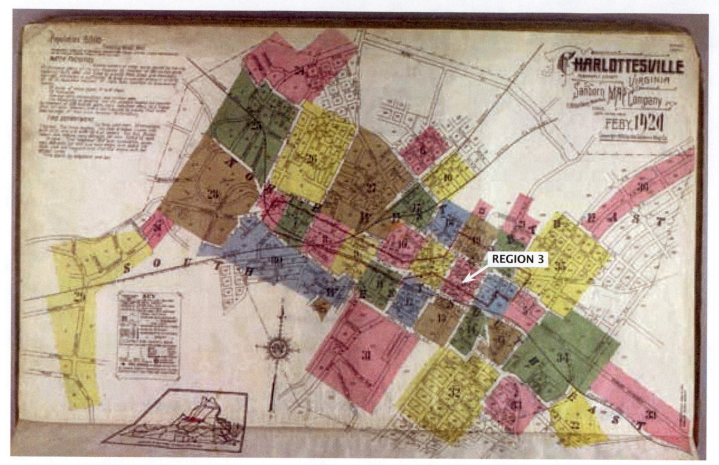

Screen shot, © Geospatial and Statistical Data Center, University of Virginia Library Web site
(http://fisher.lib.virginia.edu), 2002. Used with permission.

Step 3   On the "How to" page, scroll to the top and click on "1920 Maps." This will take you to a page listing options for viewing the 1920 atlas.

Step 4   Under "Find by map index," click on "View Full-sized Map Index" to go to a clickable map index of the city.

The full-size version will not fit on your monitor, so to help you navigate, a small version of the index is reproduced above.

Using the index, you can link to the individual, detailed sheet maps of the city.

Step 5   Navigate to the downtown region of Charlottesville, an area depicted in pink on the index, where Main and Market streets parallel each other. This is labeled region "3."

Step 6   Click on the "3" to link from the index to Sheet 3, the detailed map for that region, via the Sheet 3 menu.

You should now be at a page that looks like the screen shot on page 127.

Step 7   To begin with a reading of the industrial activity of downtown, scroll down to "Sheet 3 Top (left page)" and click on "Full Image." This image will take some time to download; however, you will find the full resolution easiest for map interpretation.

Step 8   Explore the detailed image of Sheet 3. Scroll down to E. Main Street, which runs horizontally across the bottom of the sheet.

Question 3: What kinds of industries exist on E. Main Street? To answer this question, use the vocabulary for interpreting industrial activity (manufacturing vs. service sector; tertiary, quarternary, quinary, etc.) that you have learned in class.

Question 4: In what types of buildings were these services located? (If you cannot remember the codes used in the Sanborns, you can reread the symbol explanation any time by clicking on the "Symbols" link at the top or bottom of the page.)

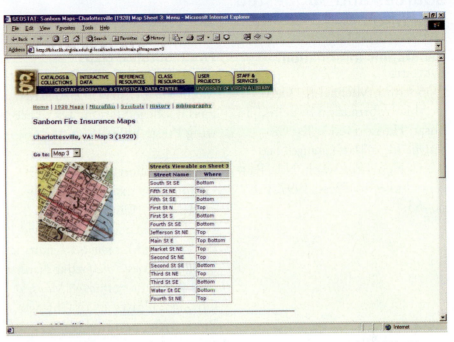

Screen shot, © Geospatial and Statistical Data Center, University of Virginia Library Web site (http://fisher.lib.virginia.edu), 2002. Used with permission.

Now begin exploring northward by scrolling slowly upward on the map.

Question 5: How does industrial activity change as one travels away from E. Main Street?

Question 6: Using the mapped information about building size and type, and comparing the lot sizes, how do you think the cultural landscape is changing?

Step 9   Scroll up to the top of the page and click on "View Bottom of Map." This will link you to the map detail for the south side of E. Main Street.

Question 7: What is the difference in industrial activity between the neighborhood north of E. Main Street (along Market Street) and the neighborhood south of E. Main Street (along Water Street)?

Question 8: How does the railroad alter the industrial geography of the city?

Step 10   Click on the "1920 Maps" link at the top or bottom of the page, and look again at the map index. This time, click on "View Mid-sized Map Index."

Question 9: Imagine that you are a traveler on the Chesapeake & Ohio Railroad. You can see on the index that the Chesapeake & Ohio line enters the city of Charlottesville from the northwest corner and exits from the southeast.

It is 1920, and you are traveling east. Charlottesville is not your destination, only a city that you are traveling through. As the train runs slowly through the city, you get a clear view of the city from your window on the south side of the train.

What do you see?

Describe the changes in the industrial landscape as you travel through Charlottesville. Use the index to guide you to the detailed sheet maps through which the Chesapeake & Ohio tracks run.

Sources and Suggested Readings

Geographic Information Systems

Goodchild, Michael F. "Geographic Information Systems," in Susan Hanson (ed.). *Ten Geographic Ideas That Changed the World.* New Brunswick, N.J.: Rutgers University Press, 1997, pp. 60–83.

Fire Insurance Maps

Oswald, Diane L. *Fire Insurance Maps: Their History and Applications.* College Station, Tex.: Lacewing Press, 1997.

Karrow, Robert, and Ronald E. Grim. "Two Examples of Thematic Maps: Civil War and Fire Insurance Maps," in David Buisseret (ed.). *From Sea Charts to Satellite Images: Interpreting North American History through Maps.* Chicago: University of Chicago Press, 1990, pp. 213–237.

CHAPTER
10

Urbanization

Vocabulary applied in this chapter
urbanization
urban sprawl
urban morphology
urban landscape

New vocabulary
data polygons
pixel
overlay

🔴 10.1
Urbanization and Population Indicators

As you have learned in class, the meaning of **urbanization** for any one place depends in part on whether the place is in a developing country or a developed country. Even more complicated is the fact that there is no agreed-upon, single definition of what level of population density constitutes an urbanized area. How many people make a city? The answer varies by country. Yet, as human geographers we can still find useful ways to make comparisons between urban regions.

This exercise explores the patterns of urbanization in the developing and developed worlds. We will compare urban and rural with data, such as household size, birth rate, and employment rate, to measure the differences between the developing country and developed country. The data for the United States (our developed country) and Mexico (our developing country) comes from the US-Mexico Demographic Data Viewer, a project of the Center for International Earth Science Information Network (CIESIN) at Columbia University.

Note: This Web site uses a particular type of network connection that may not be acceptable to all networks. If you are unable to work on this Web exercise from your computer, try accessing the site from your campus network or another type of network connection available to you.

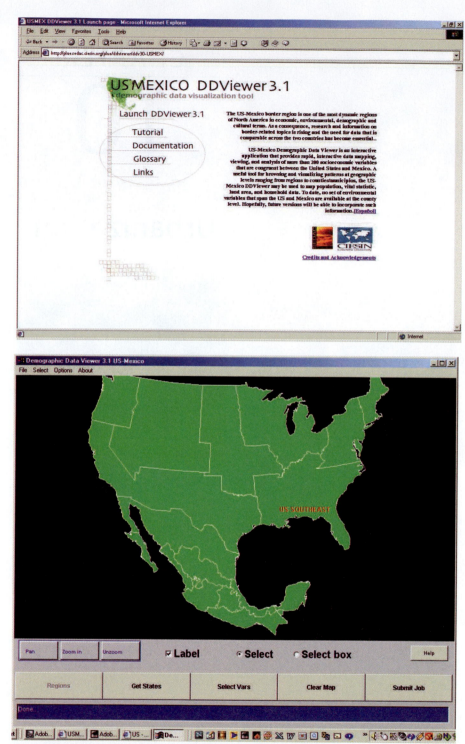

Screen shots, Center for International Earth Science Information Network (CIESIN) Web site, Columbia University, 2000. US-Mexico DDViewer, 3.1. Palisades, NY: CIESIN, Columbia University. Used with permission.

Step 1 Launch your browser and go to **http://plue.sedac. ciesin.org/plue/ddviewer/ ddv30-USMEX/**.

This should take you to the introduction to the Demographic Data Viewer, as shown in the figure at the top of page 130.

Step 2   Click on "Launch DDViewer 3.1." This may take a few moments to load.

Step 3   Under the Options menu, select the largest window size that will fit on your screen, and adjust the window so that you have a large, green map of the U.S. and Mexico with the lower buttons still visible below the map, as in the figure at the bottom of page 130.

The enumeration areas in the map are symbolized by white lines; because this map is connected to a digital database, the enumeration units are called, in GIS vocabulary, **data polygons.** In the map at which you are looking, the data polygons are broad regional divisions. As you move your cursor across the green map, you will notice that floating labels provide you with the names of these regional divisions.

The first task is to find out where the urban regions are in the United States and Mexico.

Step 4   Click on each regional division so that it turns orange, indicating that the area has been selected. When the entire map is orange, click on the bottom button "Select Vars" to select your data variables.

Step 5   In the pop-up menu, you will see several buttons indicating which data sets are available for these regional divisions. With the population data set selected, scroll to the bottom of the data set and highlight the variable name "pcturb_pop," the variable for percent of population that lives in an urban area. Next, click "Close."

Step 6   Click the bottom right-hand button, "Submit Job." The map will redraw the region polygons according to your selected data set.

Question 1: What do you see? Describe the geographical pattern of urban population density.

These regions are too broad for us to get much of a picture of geographical differences. To fix that, we need to make the data polygons smaller, so that greater data detail can be seen in the map.

Step 7   Click the bottom button, "Select New Area." This should take you back to your green map.

Step 8   This time, select the regions by dragging a box around the whole map, selecting all the regions at one time. First, click on the radio button, "Select Box." This changes your cursor to crosshairs. Next, drag a box around the whole map until it is completely orange. Click on the bottom button, "Get States."

Your green map should now be redrawn with the smaller, state-level data polygons. We want to look at the data at the county level, however, so we have one more level to go.

Step 9   Click on "Select Box" again, drag a box around the whole map, and click the button "Get Counties."

Your map should now be redrawn to show county-level divisions for both countries.

Step 10   Remap the U.S. and Mexico to show Percent Urban Population by county. Select the counties with your selection box, click on "Select Vars," set the variable to "pcturb_pop," close the variables window, and click "Submit Job." Be patient: This is a large data set, and it will take a few minutes to process and map.

When the map displays, you will notice immediately that it is difficult to see the color in the smaller counties because the color of the county border is interfering with the interior color.

Step 11   Fix the color problem by going to the Options menu, selecting "Output," and deselecting the check box next to "Boundary color." Next, click on "close."

Question 2: What is the geographical pattern of urban population density that you see?

Step 12   As you move your cursor around the map, the floating labels will give you the exact percentage for each state. Use the "Pan," "Zoom in," and "Zoom out" functions to explore different parts of the map.

Question 3: How would you describe the urban population density in the U.S.-Mexico border region? How do the coasts compare to the interior?

Comparing Developed and Developing

In this section, we will expand our use of the DDViewer to look at some of the population indicators related to urbanization.

Question 4: Think about what you have already learned about the general differences between the urban populations of developing countries and the urban populations of developed countries. What do you expect the difference in rate of urbanization to be? In household size? Birth rate? Explain why you expect to find each of these differences in the data.

Now let's test the assumptions concerning developed and developing countries to see if they hold true for a comparison of Los Angeles and Mexico City.

Step 13   Return to the US-Mexico DDViewer and click on either "Select New Area" or "Regions" (depending on where

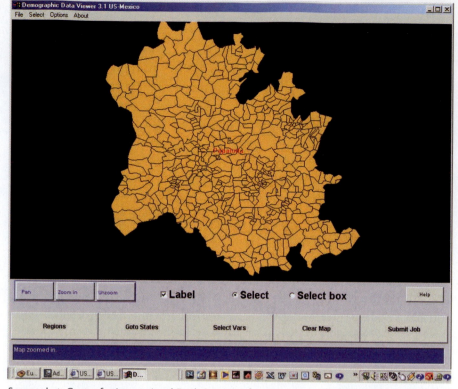

Screen shot, Center for International Earth Science Information Network (CIESIN) Web site, Columbia University, 2000. US-Mexico DDViewer, 3.1. Palisades, N.Y.: CIESIN, Columbia University. Used with permission.

you left off in the last section) so you are back to the initial green map of the U.S. and Mexico. If the map does not show polygons along regional boundaries, reset the map to display the regional divisions.

Step 14   Start with Mexico City. Select the region "Mex-Central" and click on "Get States." This will zoom you to the states around Mexico City.

Step 15   Select all seven states for this region and click on "Get Counties" to bring the polygons to the county level.

Step 16   Select the counties, and select "pcturb_pop" from the Population menu as you did before.

Step 17   Before closing the "Select Variables" window, add

some variables from the Households data set. Click the "Households" button, and browse the data sets available there.

Because we want to compare the proportion of small and large household sizes, select one or two data variables that will give you a good comparison. Remember what you learned in chapter 3: because this is a choropleth map, enumeration units should show proportional, not count, data.

Step 18   When you have some interesting household variables selected, close the Variables window and click "Submit Job."

Question 5: Where are the households the smallest? The largest?

Question 6: Explore the different spatial patterns portrayed by household size and percent urban. Is there a correlation between the two data sets? Do highly urbanized counties tend to have large households, small households, or is there no pattern?

Question 7: Compare your findings to your expectations in Question 4. Does the data support the theory for a developing country? Why or why not? You can use the "Statistical Info" button in the lower left-hand corner to help you interpret what you see.

Step 19 Now make the same comparisons for birth rate. In the Variables window, select both the 1990 and 1994 data sets.

Question 8: What is the difference in the spatial pattern of high birth rates from 1990 to 1994?

Question 9: How does this compare with your expectations from Question 4?

When you have analyzed the population indicators, you can close the DDViewer and quit the browser for the next exercises.

10.2 Interpreting Urbanization from Satellite Imagery

As you learned in your textbook, although the world is becoming increasingly urbanized, different countries have different proportions of their populations residing in urban areas. The proportion of urbanized population in any one country influences the degree to which that country is affected by **urban sprawl.**

How do we measure that? What does urban sprawl look like? And how is it affecting not only the neighboring regions into which it spreads, but continents as a whole?

The difficulty of mapping urban sprawl at a regional or continental level has been an ongoing issue for geographers. Landsat TM, AVHRR, and SPOT imagery (see exercises in chapter 8) are useful for interpretation of regional trends in land cover, but not necessarily for interpretation of differences between the built and nonbuilt environments. In part, this is due to the scale problem: Landsat TM and AVHRR imagery do not offer the level of resolution necessary for classifying rural vs. urban areas. The higher resolutions of SPOT and other imagery sources can be used to map and monitor land transformations at the local level, but the large file sizes for such detailed digital imagery make it all but impossible to work outward to regional, much less global, land cover interpretation.

In 2001, scientists at NASA Goddard Space Flight Center, under the direction of biologist Marc Imhoff, created a global satellite image that shows the locations of light on the surface of the earth. As you can see in the image on page 134, the clustering of light shows the density and intensity of urbanization.

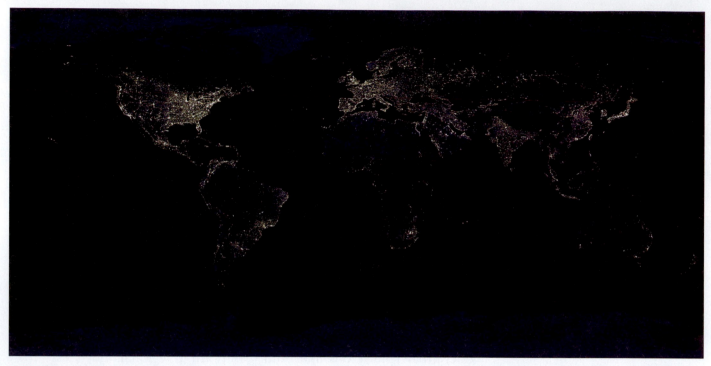

"Global City Lights" image by Robert Simmon and Craig Mayhew, Distributed Active Archive Center (DAAC), Goddard Space Flight Center, 2000. Used with permission of Robert Simmon.

Dot distribution map of world population density (1 dot equals 500,000 people). Enumeration unit: Country. Population data and base map data sources: ESRI 1999.

Question 1: Compare the urbanization pattern in the NASA image with the dot distribution map of population density below. What is the difference in the information in each of the maps?

Question 2: From what you have learned about dot distribution maps in chapter 3, what is the difference in the locational information of the yellow "lights" in the top image, versus the yellow dots in the bottom image?

Question 3: Taken together, what can these two maps tell us about the relationship between lights on the earth and population density? Is this relationship the same across the globe?

The urbanization image was created with a satellite network called the Operational Linescan System (OLS). The satellite sensors in the OLS measure lunar reflection on cloud surfaces, data gathered for use by the Air Force. This reflectance is recorded as a grid of reflectance values in a satellite image. Each of the cells in the grid of a satellite image is called a **pixel**, and each pixel carries information about reflectance intensity for that particular location.

When lunar reflection is at its lowest during the new moon, the sensors are able to collect light emitted from the earth's surface. It is this new moon imagery that Imhoff used to create his global visualization of urbanization.

Using the Image to Assess the Effects of Sprawl

Creating an image of urbanization is one thing, but how can it be used to get a visual understanding of the effects of sprawl? Imhoff's solution was to apply the simple concept of **overlay** from GIS: overlay the global lights data with other georeferenced biological data, such as soil types and vegetation.

To do this, Imhoff's team had to convert the lights image into a GIS data layer by translating intensity of light levels into urban classification levels. Because the nature and extent of the urbanized population varies by region across the globe, the team had to figure out how to convert light intensity into population density for each region separately.

For the United States, they combined the lights image with an overlay of census data showing population density. By comparing the two data layers, they were able to then classify each pixel in the lights image according to whether the intensity indicated urban, peri-urban (small towns and agricultural areas), or non-urban (10 people per square mile or less).

The newly created urban classification data layer was then compared to a data layer classified according to soil limiting factors. This data layer was derived from the soil types depicted in the UNFAO Digital Soils Map of the World.

By converting the soil types to soil fertility, the scientists were able to create a soil limitations data layer. The portion of this data for the United States is shown on page 136.

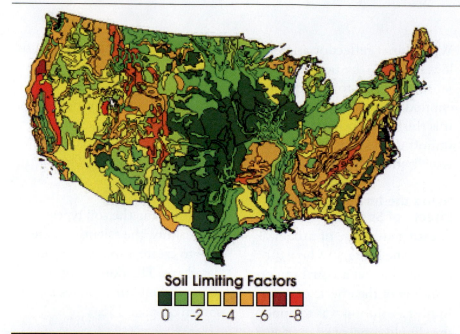

Soil Limiting Factors

0 -2 -4 -6 -8

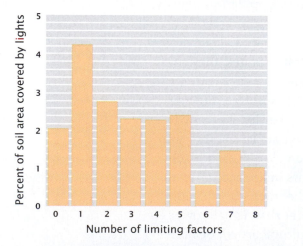

Number of limiting factors

"Soils Map of the United States" and "Percent of Soil Area Covered by Lights in the United States," by Dr. Marc L. Imhoff, Biospheric Sciences Branch, Goddard Space Flight Center. Bar graph adapted from Figure 3.3 by Dr. Marc L. Imhoff. In Imhoff, Marc L., et al., "Assessing the Impact of Urban Sprawl on Soil Resources in the United States Using Nighttime 'City Lights' Satellite Images and Digital Soils Maps," *Land Use History of North America*, United States Geological Survey (http://biology.usgs.gov/luhna/chap3.html). Used with permission of Dr. Marc L. Imhoff.

In the map on the left, the green areas show a low number of soil-limiting factors, and the red areas show a high number of soil-limiting factors.

Imhoff's team then compared the "Soil Limiting Factors" map to the "City Lights" map to find out how the two types of information compared. Their results are in the bar graph below left.

Question 4: Compare the information in the bar graph to the Soil Limiting Factors map. Which type of soil was found to be most covered by "lights," or most urbanized? Which type of soil is the second-most urbanized?

Question 5: Which type of soil was found to be least urbanized?

Question 6: Which soil types are as a result most threatened by the impact of sprawl? Why do you think this is?

Sprawl also affects soil biodiversity. In its report, Imhoff's team noted: "Our results indicate that four soil types, as classified in the UNFAO system in the United States, may be in danger of disappearing under urban/suburban structures."

Question 6: Compare the "Soil Limiting Factors" map with the inset of the "City Lights" map. Which states seem most likely to be losing soil biodiversity to sprawl?

10.3
Planning Urbanization

Mapping **urban morphology** is an ancient geographical problem. A city's structure is defined not only by its layout of streets and the distribution of land use, but also by differences in architecture, roads, views, and infrastructure. Like other large-scale mapping, urban cartography must take into account not one but all of the possible cartographic perspectives: plan, oblique, and profile, each of which provides a different perspective of morphology.

The challenge of mapping urban morphology is even greater when the task is to present a vision of a city that does not yet exist, the potential but as yet invisible **urban landscape.** A good plan should include the streets, buildings, and parks, whether public or private, at correct scale relative to each other, and in such a way that the viewer or reader can imagine exactly how the city will be.

How do you visualize the plan of a city so that it will be accessible to a wide audience, before that city is built? In this section, we will explore an early twentieth century solution to this problem.

In 1909, the Commercial Club of Chicago presented a *Plan of Chicago* by the architects Daniel H. Burnham and Edward H. Bennett, with illustrative plates primarily by Jules Guerin, Fernand Janin, and Bennett. The *Plan* was a sweeping study of proposed changes to the parks, transportation, and neighborhood systems of the city of Chicago, using European cities, such as Paris, Vienna, and Rome, as models.

Below is a conventional map from the *Plan*, showing existing and proposed changes to the region from Grant Park south to the proposed Civic Center. The map is oriented with south at the top, and Lake Michigan and the Grant Park piers at the bottom. Proposed new or widened streets are shown in dark red, proposed parks are shown in light green, and existing parks are dark green. By using color, the design team gives the reader a clear overall sense of the changes proposed to the structure of the city.

But knowing how the structure will be different doesn't give much of an idea of what it will be like to *live* in this new Chicago. Burnham and Bennett wanted to address the latter idea as well, to convey a sense of the new city plan the way one gets a sense of the city while experiencing it directly.

Detail from plate 86, "Chicago. Plan of the Street and Boulevard System Present and Proposed," in *Plan of Chicago* by Daniel H. Burnham and Edward H. Bennett, Commercial Club of Chicago, 1909.

To enhance the conventional maps, Jules Guerin created aerial and perspective views, landscape sketches, and architectural profiles, creating a massive vision for the city published in a volume of maps similar in format to an atlas. The *Plan of Chicago* was a model of visualization technique before the advent of digital visualization and virtual urban reality. Compare the map on page 137 to the images on these two pages. Each illustration represents portions of the same axis from Grant Park to the proposed Civic Center.

Question 1: How does Guerin use scale to convey a sense of the imagined city?

Question 2: How does he use direction and perspective?

Question 3: What other techniques does Guerin use to bring the proposed urban plan to life? Which elements of the maps are particularly good for conveying a sense of the urban landscape?

Question 4: What elements of the imagined city do you think are unobtainable through maps and views?

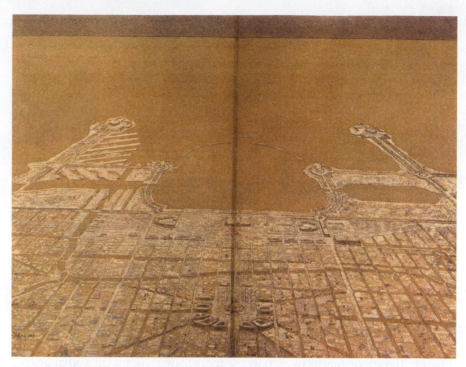

Detail from plate 137, "Chicago. View of the Proposed Development in the Center of the City, from Twenty-Second Street to Chicago Avenue, Looking Towards the East over the Civic Center to Grant Park and Lake Michigan," in *Plan of Chicago*.

Detail from plate 127, "Chicago. Bird's-Eye View at Night of Grant Park, the Facade of the City, the Proposed Harbor, and the Lagoons of the Proposed Park on the South Shore," in *Plan of Chicago*.

Detail from plate 128, "Chicago. Proposed Plaza on Michigan Avenue West of the Field Museum of Natural History in Grant Park, Looking East from the Corner of Jackson Boulevard," in *Plan of Chicago*.

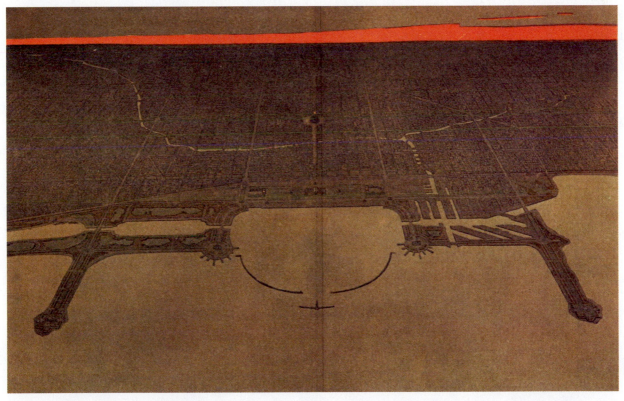

Detail from plate 87, "Chicago. View Looking West over the City, Showing the Proposed Civic Center, the Grand Axis, Grant Park, and the Harbor," in *Plan of Chicago*.

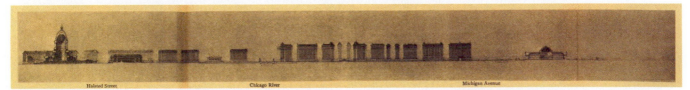

"Chicago. Section Looking North, Taken through Proposed Grand Axis of the City, Showing the Civic Center and Grant Park," plate 126 in *Plan of Chicago*.

Sources and Suggested Readings

Urbanization and Imagery

Imhoff, Mark L., et al. "Assessing the Impact of Urban Sprawl on Soil Resources in the United States Using Nighttime 'City Lights' Satellite Images and Digital Soils Maps," in United States Geological Survey. *Land Use History of North America.* http://biology.usgs.gov/luhna/chap3.html.

Weier, John. "Bright Lights, Big City." NASA Earth Observatory, October 10, 2000. http://earthobservatory.nasa.gov/Study/Lights/

Planning Urbanization

Burnham, Daniel H., and Edward H. Bennett. *Plan of Chicago.* 1909.

Danzer, Gerald A. "The Plan of Chicago by Daniel H. Burnham and Edward H. Bennett: Cartographic and Historical Perspectives," in David Buisseret (ed.). *Envisioning the City: Six Studies in Urban Cartography.* Chicago: University of Chicago Press, 1998, pp. 144–73.

CHAPTER
11

The Urban Mosaic

Vocabulary applied in this chapter
models of the city
Central Business District
   (CBD)
commuting
empirical observation
participant observation
neighborhoods

⊕ 11.1
Urban Models in Practice

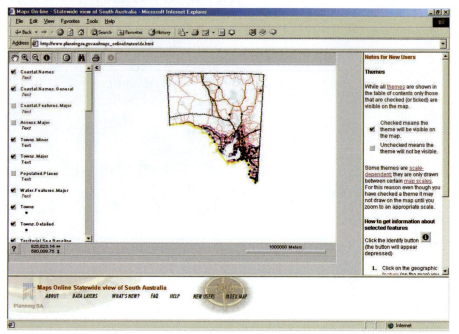

Screen shots above, Planning SA Web site (www.atlas.sa.gov.au), 2002. © Planning SA. Used with permission.

As you have learned, geographers have theorized several structural **models of the city:** the **concentric zone model,** the **sector model,** the **multiple nuclei model,** as well as models for South Africa, Latin America, and the former Soviet Union. How do these models compare to the rest of the world? This exercise explores the structure of Adelaide, using its Web GIS "Maps On-line."

Step 1   Launch your browser and go to the Maps On-line Web page at **www.planning.sa.gov.au/ maps_online/start.html.**

This will take you to the gateway to the interactive Web map, as in the image above, right. Note the location of Adelaide, the city to which you are headed in this exercise.

Step 2   From the left column, select "Entire State" to launch the interactive map. This will take you to a window giving you an overview map of the whole state of South Australia, as shown in the bottom screen shot at the right.

Step 3   In the list of available data layers on the left, scroll down and check the box next to "Roads. Detailed." The roads will not appear on the map at this scale, but will appear later when you are analyzing a view with a larger scale.

Step 4   Zoom in toward the city of Adelaide, the largest city on the coast in the Gulf of St. Vincent. To zoom, click on the magnifying glass tool in the menu bar across the top of the map, click the cursor, and drag a box over the area of Adelaide on the map.

Step 5   Continue to zoom in on Adelaide until you begin to detect the pattern of streets and blocks. If you lose part of the city from the window, use the Hand tool in the toolbar to pan back

to the part of the map you wish to explore. Click the Hand tool, then click and drag the map to recenter it.

Step 6   When you have reached a scale at which you can read the street pattern of Adelaide (approximately at the point when the bar scale is indicating "10,000 meters" across), add the zoning data to your map. Scroll down the data layers legend to the left of the map, and turn on the "Generalised Zoning" layer by clicking on the box with your

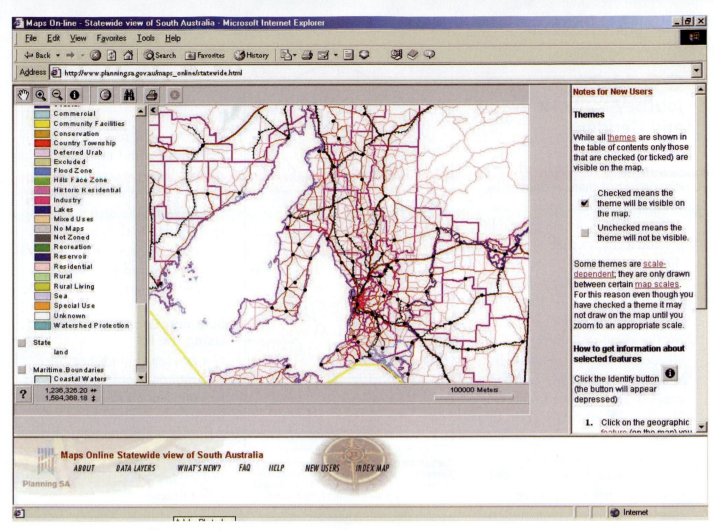

cursor. The map will automatically redraw with zoning data added.

Question 1: Where are the major recreational areas located with respect to the downtown? Industrial areas?

Step 7 Continue to zoom in toward the center of Adelaide, observing the detail of land use in this city.

Question 2: The Torrens River runs generally east-west through Adelaide. What is the zoning along this river?

Step 8 On a separate sheet of paper, make a sketch showing general regions of the residential, recreational, industrial, and commercial districts.

Question 3: Compare your sketch to the urban models from your textbook. Which model does it most closely resemble?

Question 4: Does Adelaide have a Central Business District (CBD)? If so, where is it located?

Question 5: From the land-use categories available in this GIS, is it possible to differentiate low- and high-income residential areas? Explain why or why not.

Is Adelaide typical for South Australia? Using the same steps as above, compare the land-use patterns of Port Lincoln, Port Augusta, and Mount Gambier.

Step 9 Make a general sketch for each of the other three cities. Compare your four sketches.

Question 6: Is there a typical model for the South Australian city? If so, what does that model include? If not, how can you account for the differences?

✺ 🔍 11.2
Journey-to-Work and Gender

In chapter 7, you explored the details of population density by comparing the spaces of where people work vs. where people live around a city. In that exercise, the focus was to think about the way that flows of population change depending on the time of day. For the purposes of clarity, the assumption was made that all commuters are the same. But what if we had looked at differences in the type of commuter? How would our image of the city change?

Urban geographers first started looking at differences in **commuting** distances between men and women. Women, it was found, tend to work closer to home than men, and so have lower average commuting distances to travel. When geographers started looking at *why* this was true, however, they found that it wasn't just a simple story about gender—other socioeconomic variables greatly affected those travel distances as well. In this exercise, we will map their findings and analyze the results for ourselves.

Note: You will need a compass to draw circles on the maps, and a calculator to do some arithmetic for this exercise.

In their 1995 book *Gender, Work, and Space*, Susan Hanson and Geraldine Pratt looked in detail at the question of commuting

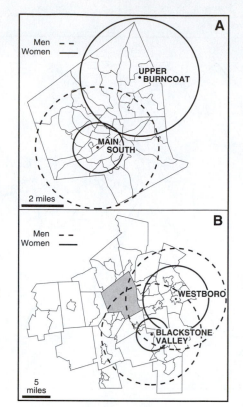

"Median Journey-to-Work Distances of Men and Women Living in Four Local Areas in Worcester, Massachusetts," from Susan Hanson and Geraldine Pratt, *Gender, Work, and Space,* 1995. Used with permission of the authors and publisher. Map by Anne Gibson.

distances and women in Worcester, Massachusetts. Using maps, Hanson and Pratt visually compared the median distance to work for women to the median distance to work for men.

The map above depicts the differences for men and women commuters in the City of Worcester (Map A), and the Worcester suburbs (Map B). Journey-to-work distances for women are shown with a solid line radius, and a dashed line indicates journey-to-work for men.

Question 1: Analyze Map A, above, using the bar scale. What do you estimate is the difference in journey-to-work distances for men and women?

As you can see in the map on the left, Hanson and Pratt found that the gap between commuting distances between men and women in the Worcester area was wider in the suburbs than in the city. How does Worcester compare to other cities, and what can be learned from such comparisons?

Mapping Commuting Radii Using the RF

Geographer Ibipo Johnston-Anumonowo looked at the commuting data for men and women in the city of Baltimore. She found that men's average distance to work is 7.8 miles, and women's average distance to work is 5.5 miles.

As in the Worcester map on the left, a visual impression of that commuting difference can be achieved by plotting the commuting distances on a map as radii of a circle, as shown in the map, opposite. The radii for the two circles were drawn from the center of the city at the "**x**".

Now, let's get a closer look at the commuting data, and compare it to other socioeconomic information. In this next section, you will map two additional variables: One-Worker vs. Two-Worker Households, and Central City Residents vs. Suburban Residents.

Step 1   Photocopy the four maps of Baltimore on page 147, so that you can work from the photocopies.

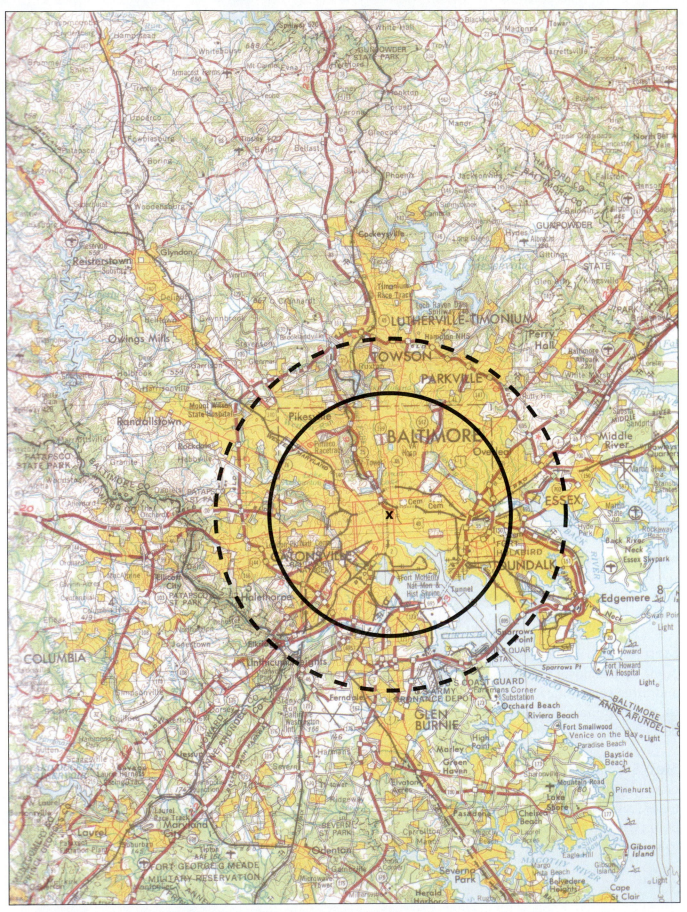

A comparison of commuting distances for women (solid circle) and men (dashed circle).
Base map is a portion of *Baltimore*, 1:250,000, U. S. Geological Survey, 1957 rev. 1978.

miles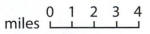

Question 2: The RF of these four maps is 1:618,000 (to brush up on RF and map scale, refer to chapter 2). Based on that RF, 1 inch on the map equals how many miles on the ground? Remember that there are 63,360 inches in one mile.

Step 2   Examine the data for One-Worker Households:

Average journey-to-work for One-Worker Households
Men: 6.9 miles
Women: 5.5 miles

This data will go on the first map on your photocopy, "Map #1: Average Commuting Distances in One-Worker Households."

Step 3   Using the RF, calculate the radius for the men's circle by translating the distance they travel on the ground (6.9 miles) to the equivalent distance on the map.

Step 4   Choose a center point for the city of Baltimore, and using your compass, draw in the travel-distance circle with a dashed line.

Step 5   Next, calculate the radius for the women's data (5.5 miles), and using the same center point, draw in the second circle using a solid line.

Step 6   For "Map #2: Average Commuting Distances in Two-Worker Households," draw two separate travel-distance circles for men and women using the following data:

Average journey-to-work for Two-Worker Households
Men: 8.3 miles
Women: 5.5 miles

Finally, to the comparison of men and women, and one- and two-worker households, we will add a third factor: whether the commuter resides in the city or the suburbs. Map #3 depicts the commuting distances of city residents for both one- and two-worker households. As before, the dashed line represents commuting distances for men, and the solid line represents commuting distances for women. One-worker households are shown with blue circles, and two-worker households are shown with green circles.

Question 3: Compare the commuting radii in Map #3. How do the commuting distances of men and women in this map compare to those of Maps #1 and #2?

Question 4: What factors do you think might have influenced these results?

Step 7   Now compare the city dweller data to the data for suburban dwellers. Using the data below, complete the circle radii for "Map #4: Average Commuting Distances from the Suburbs: One-Worker and Two-Worker Compared."

Suburban Residents, Single-Worker Households
Men: 7.9 miles
Women: 7.3 miles

Suburban Residents, Two-Worker Households
Men: 9.3 miles
Women: 6.6 miles

Since this is suburban data, you will need to choose one of the suburban places outside of Baltimore as the new center point for your circles. As in Map #3, use two different colors to differentiate the circles for one-worker and two-worker households.

Question 5: How does Map #4 differ from Map #3? Does this fit your speculation from question 4?

Question 6: Compare the four Baltimore maps to the maps of Worcester on page 144. Which factor, dwelling location or number of workers per household, seems to have the greatest impact on the commuting distances of men and women?

Map #1
Average Commuting Distances
in One-Worker Households

Map #2
Average Commuting Distances
in Two-Worker Households

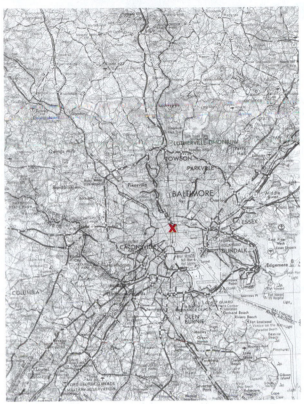

Map #3
Average Commuting Distances from the City:
One-Worker and Two-Worker Compared

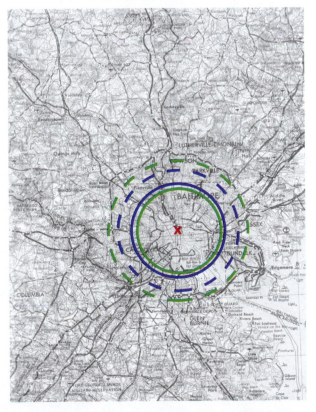

Map #4
Average Commuting Distances from the Suburbs:
One-Worker and Two-Worker Compared

⊘ 11.3
The Social Structure of the City

In the 1870s, Charles Booth decided to investigate the social conditions of the London poor through a detailed cartographic inquiry into the spatial distribution of those conditions. Like many of his contemporaries, Booth was troubled not only by the living conditions in London's poor neighborhoods, but the limited way in which these conditions were understood.

Dedicated to the combined use of different social scientific methods for the most comprehensive picture of the city, he and his staff set out on what would become more than three decades of map production depicting the interrelationships of demographic statistics for most of the city of London. The results of this sweeping, landmark study were the publications between 1889 and 1899 of three series of studies entitled *Life and Labour in London*: the *Poverty Series*, the *Industry Series*, and the *Religious Series*.

In their quest to create a social picture of what was then the largest city in the world, Booth and his staff made use of social data sets available to the public supplemented by their own **empirical observations** of **neighborhoods** and streets, and **participant observation.** During the early years of the project, these disparate data sources were combined primarily as written narratives of buildings and neighborhoods. Over time, Booth became more involved in representing these narratives with maps and statistical tables.

Although Booth was not the first to measure London's social conditions through mapping, the maps of *Life and Labour* achieved a level of cartographic detail never before accomplished in urban cartography. Today, the maps stand as a model of the possibilities for social cartography, both in scale and scope.

In this exercise, we will follow Booth's interpretation of the fabric of urban life through distribution mapping as it evolved over time: to record the characteristics of the neighborhoods themselves, to analyze the interrelationships of social forces within those neighborhoods, and to analyze the urban social structure as a whole. To explore how both the city and Booth changed over time, we will look at portions of two of Booth's major maps.

Step 1   To go to the
first map, launch your browser
and navigate to **www.umich.edu/
~risotto/home.html.**

This site, created by Sabiha
Ahmad and David Wayne
Thomas, is an accessible online
version of the first major map
to come out of Booth's project
in 1889: the "Descriptive Map of
London Poverty." The "Descriptive Map" depicts street-scale clas-
sifications of wealth and poverty
on a street map of London at a
scale of 6 inches to a mile.

Step 2   Click on "Map" from
the graphic on the main page to
go to the clickable map version,
titled "Imagemap." You will see
the 1889 map in full, with individual squares that can be clicked
on to zoom in along the map (as
you did to explore Sukula's map
in chapter 3). The Thames River
cuts a curvy, pale swath through
the city, separating the southeast
neighborhoods from the northwest neighborhoods.

Step 3   Although the grid does
not show any reference numbers
or letters, Booth's map does have
its own index, which runs horizontally from A to M, and vertically from 1 to 12. Click on the
upper left corner grid cell to see
the layout of the grid: the Web
site will link you to the beginning
of the grid at A/1-2, as in the
bottom image on the right.

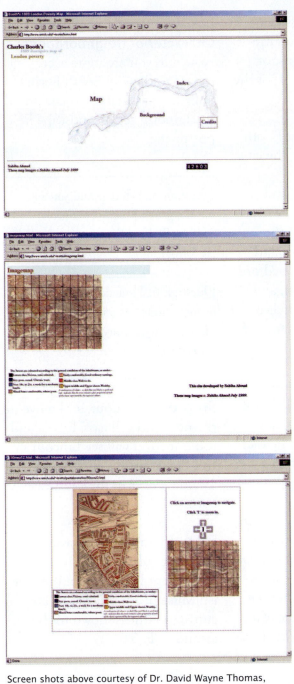

Screen shots above courtesy of Dr. David Wayne Thomas,
Department of English, University of Michigan.

Step 4   Take a moment to look
at the legend, which notes that
"The Streets are coloured according to the general conditions of
the inhabitants...."

*Question 1: What kind of geographi-
cal data is this legend portraying? Is
it qualitatively different social data,
or is the data ranked?*

*Question 2: How are these character-
istics conveyed by the colors Booth
has chosen for the map?*

From the zoomed-in view, you will see that from here you can also pan in four directions, or Zoom In by clicking the "I" at the center of the pan arrows.

Let's jump in and look at Booth's data right in the thick of the city.

Step 5

Using the pan and zoom features of this interactive map, go to cell E/5-6, on the north side of the Thames, and zoom in. You should see Bedford Square in the top portion, and Piccadilly Circus and Trafalgar Square in the bottom portion of the cell.

Question 3: Examine Booth's use of color to depict categories of social status in the neighborhoods. In what way is this technique similar to a choropleth map?

Question 4: As you learned in Chapter 3, in a conventional choropleth map, data is depicted by enumeration unit. In this map, the units are considered "units of analysis"; the information behind the units is not of a strictly numerical nature. What is the unit of analysis used in the map?

Interpreting the Social Character of Neighborhoods

Now that you are more accustomed to the way Booth is symbolizing information, focus on the London social landscape he is portraying.

Question 5: What would you say are the general social conditions around Bedford Square and the British Museum?

Question 6: How do these social conditions change as one moves west? How do they change moving south into Soho?

Step 6

Zoom out, and move to the next grid cell east, F/5-6, and zoom in. You will see Lincoln's Inn Fields at the center and the Thames River in the southeast corner.

Step 7

Imagine that you have just walked east down New Oxford Street from the British Museum, and now you are cutting southeast down to the Strand in F/5-6.

Question 7: How is the social landscape changing? To answer this question, compare for example:

- *Types of workplaces and institutions (public buildings? churches? factories?)*
- *Types of transportation, quality and level of roads*
- *Economic class (pattern of poverty and wealth)*
- *Size and quantity of residences*

Question 8: Does it appear that changes in social conditions are gradual in these neighborhoods or abrupt? What kind of pattern in the map indicates this type of change?

Question 9: Try to imagine yourself in each of these neighborhoods. What is it like? For example, which of these neighborhoods are:

noisy or quiet?
crowded or deserted?
at which times of day?

Step 8

Now explore around the rest of the city using your pan and zoom arrows again, to think about and answer the following questions.

Question 10: How do the street structures for the poor classes differ from those of the wealthier classes? Consider the width of the streets, the existence of cul-de-sacs or through streets, and the distribution of alleys and avenues.

Question 11: What is the relationship of the river to the residential pattern? The parks?

Question 12: Are there physical barriers that separate the poor sections from the rest of the city?

A Decade of Change

A decade after the "Descriptive Map" was released, Booth revised and expanded it for the *Religious Series*, indicating changes to the city structure (demolitions and new construction) since 1889, making corrections from the field notes, and supplementing the map with other sketch maps and tables.

The revised map also categorized streets by color according to social conditions, as in the 1889 map. In the new map, however, Booth added church parishes because he wanted to look for the relationship between church parish and poverty.

On page 153 is "The Inner West," one of the sheets of the 1899 map. This portion shows the London neighborhoods south of Hyde Park; Kensington Station can be seen to the left of the center, south of Cromwell Road in Brompton.

This corresponds to the online 1889 map as B–D horizontally, and 7–8 vertically:

	B	C	D
7			
8			

As is clearly shown by the preponderance of yellow, this region of Victorian London was largely wealthy. Part of the significance of Booth's works, however, was the illumination of pockets of poverty and lower income in neighborhoods perceived to be entirely high income.

Step 9   Look back at your online 1889 map and find the corresponding map area according to the diagram above. This region of the map corresponds to three grid cells on the online map.

Question 13: In the 1899 revised map, how does the neighborhood in West Brompton (in St. Luke's parish) compare to that of Wilton Place (in St. Paul's parish)?

Question 14: Compare these same two neighborhoods in your online 1889 map. Have they changed? Why or why not?

Question 15: For both maps, compare the spatial distribution of the "working class" colors to that of the "well-off." Is one income class more densely clustered than the other? How do they differ?

Question 16: Do neighborhoods exist that show both black and red categories in the same block? Which neighborhoods indicate the largest gap between rich and poor?

Question 17: Has each of the income categories changed its distribution pattern? Is there one that has remained relatively similar in its spatial distribution in the city?

Although we can learn a great deal about Victorian London from these maps alone, it is not entirely fair to judge them in this way, as Booth never intended for them to stand alone. In the *Religious Series*, Booth used the color maps in conjunction with sketch maps, statistical tables, and personal observations in his goal to interpret the influence of church districts on London social conditions.

Detail from "Map M—The Inner West" by Charles Booth. In *Life and Labour of the People in London. Third Series: Religious Influences. Vol. 3: The City of London and the West End*, London: Macmillan and Co., 1902.

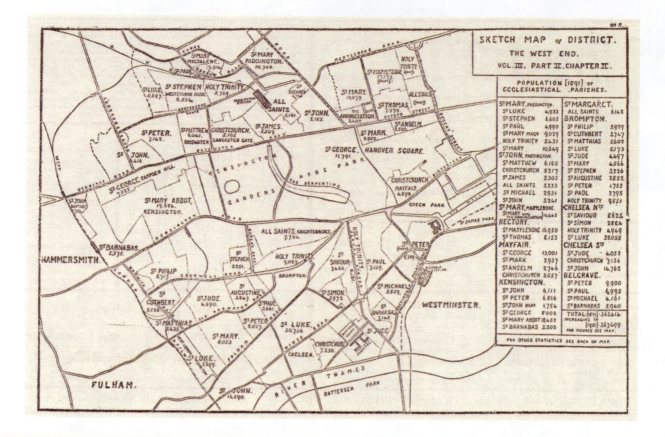

STATISTICS *bearing on the* AREA INCLUDED IN SKETCH MAP NO. II. *Described in Chapter II.* (Vol. III., Part II.).

CENSUS STATISTICS.

Showing Increase or Decrease of Population.

	POPULATION IN			**Increase** *or Decrease.*	
1881.	**1891.**	**1896.**	**1901.**	1881–1891.	1891–1901.
358,647	357,496	367,405	364,133	·29 °/₀	1·86 °/₀

Density of Population.		*Age and Sex in 1891.*			
1891.	**1901.**	AGE.	Males.	Females.	Together.
PERSONS PER ACRE. 90·5	92·6	Under 5 years	15,236	15,373	30,609
		5 & under 15 yrs	26,446	27,654	54,100
INHABITED HOUSES. 46,003	43,169	— 20 ,,	14,140	20,745	34,885
		— 25 ,,	16,285	27,946	44,231
PERSONS PER HOUSE.		— 35 ,,	27,357	44,205	71,562
7·8	8·4	— 45 ,,	19,944	28,179	48,123
		— 55 ,,	14,275	19,915	34,190
NUMBER OF ACRES.		— 65 ,,	8,865	12,946	21,811
3,948		65 and over	6,812	11,173	17,985
		Totals ...	149,360	208,136	357,496

NOTE.—The district includes MAYFAIR, and part of the BELGRAVE Registration sub-district of St. George, Hanover Square, CHELSEA, BROMPTON, the southern part of KENSINGTON TOWN, a detached portion of ST. MARGARET WESTMINSTER, ST. JOHN and ST. MARY PADDINGTON (except the Ecclesiastical parishes of St. Saviour and St. Peter), and the RECTORY, CAVENDISH SQUARE, and ST. MARY sub-districts of Marylebone (except the ecclesiastical parishes of St. Mark and St. Luke). In these figures the whole of Belgrave is included as well as the four ecclesiastical parishes mentioned above. Cavendish Square (now combined with All Souls) is omitted, as well as the detached parts of St. Margaret and Kensington Town. Owing to these omissions it is probable that the whole area is less crowded than here indicated. For Special Family Enumeration see Appendix.

SPECIAL ENUMERATION FOR THIS INQUIRY (1891).

Sex, Birthplace and Industrial Status of Heads of Families.

SEX.		BIRTHPLACE.		INDUSTRIAL STATUS.			TOTAL HEADS.
Male.	Female.	In London.	Out of London.	Employers	Employees	Neither.	
56,531 71 °/₀	22,745 29 °/₀	29,245 37 °/₀	50,031 63 °/₀	9,801 12 °/₀	47,412 60 °/₀	22,063 28 °/₀	79,276 100 °/₀

Constitution of Families.

HEADS.	Others Occupied.	Unoccupied.	Servants.	TOTAL IN FAMILIES.
79,276 (1·0)	67,100 (·85)	139,879 (1·76)	52,395 (·66)	338,650 (4·27)

SOCIAL CLASSIFICATION according to *Rooms Occupied* or *Servants Kept.*

	PERSONS.	PER CENT.	
4 or more persons to a room	13,215	3·7	Crowded
3 & under 4 ,, ,,	22,734	6·4	25·3 °/₀
2 & 3 ,,	54,426	15·2	
1 & 2 ,,	70,168	19·6	
Less than 1 person to a room	12,881	3·6	
Occupying more than 4 rooms	44,132	12·3	
4 or more persons to 1 servant	18,739	5·3	Not Crowded
Less than 4 persons to 1 servant & 4 to 7 persons to 2 servants	16,945	4·7	74·7 °/₀
All others with 2 or more servants	33,015	9·2	
Servants in families	52,395	14·7	
Inmates of Institutions (including servants)	18,846	5·3	
Total	357,496	100	

Living in Poverty (as estimated in 1889)	17·4 %}	100 °/₀
,, in Comfort (,, ,,)	82·6 %}	

"Sketch Map of District, the West End" (top), and "Statistics Bearing on the Area Included in Sketch Map No. II" (bottom), from *Life and Labour of the People in London. Third Series: Religious Influences. Vol. 3: The City of London and the West End.*

For example, on page 154 is the sketch and table for the West End.

Question 18: How do the table and sketch map alter your impression of the neighborhoods of the "Inner West," if at all?

Question 19: Do you think Booth could have incorporated the information in the table into the map? Why or why not?

Tip for exploring: If you have two computers side by side (for example, if you are working on this exercise in a computer lab), you can do the steps and questions for "A Decade of Change" completely online, because the 1899 map is also available on an interactive Web site. It will not work on one computer because you need to be able to easily look at both maps side by side, at the same time.

To do the exercise completely online, launch your second browser and navigate to the Charles Booth Online Archive at **http://booth.lse.ac.uk,** shown top, right.

At the bullet selection "Poverty Maps of London," click "Search" and select "Wards in 2000."

From the "Browse by Borough" drop-down menu, choose "Kensington and Chelsea" and click "Go."

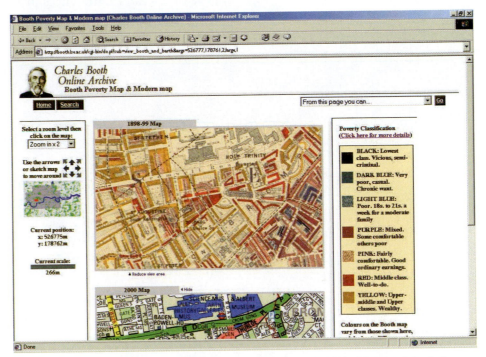

Gateway page of the Charles Booth Online Archive, 2002. © London School of Economics and Political Science. Used with permission.

Focus on Brompton from the Charles Booth Online Archive Web site, 2002. © London School of Economics and Political Science. Used with permission.

From the list of Wards, select "Brompton, Kensington and Chelsea," and click the top option, "View on Map."

This will bring you to the Brompton portion of the 1899 map, aligned about exactly with the portion reprinted in this book on page 153, as in the image above.

From here, continue with the exercise from Step 9.

Sources and Suggested Readings

Online Planning

Online Planning Journal.
**www.casa.ucl.ac.uk/planning/
olp.htm**

Journey-to-Work

Hanson, Susan, and Geraldine Pratt. *Gender, Work, and Space.* N.Y.: Routledge, 1995.

Johnston-Anumonwo, Ibipo. "The Influence of Household Type on Gender Differences in Work Trip Distance." *Professional Geographer* 44, No. 2 (May 1992): 161–69.

Social Structure of the City

Booth, Charles. *Life and Labour of the People in London. Third Series: Religious Influences. Vol. 3: The City of London and the West End.* N.Y.: Macmillan, 1902.

"Charles Booth's Descriptive Map of London Poverty 1889." [map] London Topographical Society Publication 130. London, 1984.

O'Day, Rosemary, and David Englander, eds. *Mr Charles Booth's Inquiry.* London: Hambledon Press, 1993.

Pfautz, Harold W. (ed.) *Charles Booth and the City.* Chicago: University of Chicago Press, 1967.

12

Globalization

Vocabulary applied in this chapter
globalization
Gross Domestic Product
  (GDP)

New vocabulary
histogram
line graph

⊕ 12.1
Making Globalization Visible

Globalization is a complex concept to grasp, much less measure or monitor. Most people agree that it is a combination of specific processual and structural shifts in economics, culture, and governance at the global level. These patterns include a shift from industrial to service economies, and from national to global markets, an increasing spread of popular culture and rising consumerism, and a widening gap between the rich and poor.

Question 1: Reflect on the globalization concepts you have learned in class and in your textbook. What other kinds of economic and cultural patterns are indicators of globalization?

Question 2: What kinds of activities are indicative of political and cultural resistance to globalizing forces?

Based on these patterns of globalizing forces and resistance to those forces, do you think it is possible to make a "map of globalization"? What would it look like?

Throughout this workbook, you have already had glimpses of globalization's visual pattern. For example, in chapter 2, you analyzed the parallel landscapes of vernacular and modern architecture in Lhasa; in chapter 4, you

compared two contrasting toponymic landscapes in Ireland; and in chapter 10, you analyzed the location of sprawl. In each of these cases, you were also looking at globalization: both the global effects of forces and structures on people, regions, and landscapes, and local resistance to those effects.

It is one thing to consider globalization as a series of case studies, with separate issues, indicators, and effects. But it is far more difficult to achieve an integrated awareness of globalization, a whole picture of globalization in our head. If we cannot look at it as a whole, how can we monitor it as a whole?

In this exercise, we will experiment with an online map application to see if the kinds of datasets available to us are useful for taking on a complex visual idea, globalization. To do this, we will return to the UNEP Web site from Exercise 7.1, to explore more deeply the capabilities of the GEO Data Portal.

Step 1   Launch your browser and return to the "GEO Data Portal Website" at **http://geodata.grid.unep.ch/**.

This will bring you to the familiar gateway page, shown below

Screen shot from GEO Data Portal Web site, UNEP 2002. Used with permission.

Step 2   For the search term, type in "trade" and click "search."

Step 3   Select "Trade - Percent of GDP" for the national level and click "continue."

Step 4   As you did in the exercise in chapter 7, check "Select all" for available years and click "continue."

Question 3: Based on what you have learned about regional globalization patterns, what type of data display for "Trade - Percent of GDP do you expect to see?

Step 5   Test your hypothesis by clicking on "Draw Map" from your list of options. Make the map window larger by resizing it to fit your monitor. Next, select the "Big Image" option, and click "Make new Map."

Step 6 From the time period drop-down menu, choose 1970 and click "Make new Map" again.

Question 4: Which countries or regions indicate the highest proportion of GDP in trade? Which countries indicate lower proportions?

Question 5: Redraw the map for the 1980 data. Is GDP in trade increasing or decreasing? For which regions or countries?

Question 6: Redraw the map for the most recent year available. Does the visual pattern fit your hypothesis from Question 3? Why or why not?

Question 7: How does the "No data" category affect the different views of the choropleth map? How does it affect your perception of the global balance of trade?

Question 8: Using the map interpretation skills you have gathered in this workbook, evaluate the generalization, scale, projection, and data classification of this interactive map. In what ways does each factor limit your interpretation of globalization trends?

Screen shot from GEO Data Portal Web site, UNEP 2002. Used with permission.

Using the Histogram
Next, let's check the graphic impression with a histogram.

A **histogram** shows the distribution of data values for one continuous variable. Rather than showing each individual variable along a single axis, as you saw with line graphs in chapter 7, a histogram divides the data into data classes, and then plots the frequency of occurrences of those data classes relative to the variable as a whole.

Step 7 From the "Advanced Tools" options, check "Display Histogram," and click on "Make New Map."

Step 8 Set the year back to 1970 and redraw the map. The map window should redraw the GDP data for that year, and on top of the map a histogram in a pop-up window should appear.

Step 9 Print the histogram pop-up using the print options on your computer.

Step 10 Repeat steps 8–9 for the years 1980 and the most recent year listed.

Question 9: Compare your three histograms. How is the proportion of GDP in trade changing? Does this support the concept of globalization? Explain why you think the histograms do or do not reflect globalizing forces.

Question 10: Compare the histograms to the map legends. For instance, is the shape of the 1970 histogram reflected in the data classification categories for the 1970 map? How are they the same or different?

Question 11: Do the histograms assist with your visual picture of GDP in trade? Why or why not?

Adding Indicators with the Line Graph

Whereas a histogram is useful for comparing relative proportions of a data set, a **line graph** is the best type of chart for showing data in a time series. By placing consecutive years along the x-axis, a line graph provides a useful visualization of the development of data over time.

In this section, you will supplement the map of GDP in trade, with line graphs of another globalization indicator.

Step 11 Close your histogram window so that only the map is showing.

Step 12 Instead of closing the map window, leave it open on the side or "minimized" if working on a PC. This will leave it available for later comparisons. You should now be back to the "select an option" Web page from Step 5, with the map window minimized at the bottom of your screen.

Step 13 Return to the gateway Web page by clicking "GEO Data Portal Home" in the upper left corner.

Step 14 From the "links" column in the upper right, click "data set list." Be patient: This produces a long list of available data sets, which may take a few moments to appear on your monitor.

Step 15 Explore the list of data variables available at this web site. Jot down the name of a data variable that you feel is a significant indicator of globalization, based on what you have learned about globalization in class and in your reading.

Question 12: Which variable did you choose as an indicator?

Question 13: What type of change would you expect to see in this variable in a developed country, over time?

Question 14: What type of change would you expect to see in this variable in a developing country, over time?

Step 16 As you did in Steps 3 and 4 above, select your data set for the "National" level, and then select all available years for the time period options, so that you have a good time series.

Step 17 Go back to your display options page and choose "Draw Graph." You will see a pop-up that defaults to a 2D-line graph of your selected data set.

Step 18 Scroll through the list of available countries and select a developing country of your choice to show in the graph. As before, make the graph window larger to fit your monitor screen, check the "Big Image" box under Layout, and click "Make new Map."

Question 15: Which country did you choose, and what type of pattern is indicated in the line graph?

Question 16: Evaluate whether the path of the line graph matches your expectations for that country.

Be careful interpreting countries with "No data" for one or more years. The absence of data for a particular year will throw off the line graph because in this data set, "no data" is represented by a numerical value in the graph, "– 9999." If there is a high occurrence of "No data" for the time period in the graph, try picking a different country.

Step 19 Close the line graph window, and make another line graph for a developed country of your choice.

Question 17: Evaluate the line graph for the developed country. What country did you choose, and what pattern is indicated in the graph?

Question 18: Is there a marked difference from the developing country? Do the two graphs suggest a core/periphery relationship, or do you think this is not a reasonable conclusion that can be drawn from the maps and graphs alone? Explain your reasoning.

Step 20   Do you recall your map of GDP in trade? Leaving the line graph window open, reactivate or maximize the trade map so it is again visible on your monitor. Arrange the line graph and map so that you can see the two together.

Step 21   Analyze what you see in the graph and map. For the developed country represented in your line graph, explore the data for that country over time in the map. Remember that you can use the "Identify" tool under "Basic Tools" to help you get exact numbers.

Question 19: Summarize the way in which these two globalization indicators compare with each other. As one indicator is increasing, is the other decreasing? Remaining constant? Increasing at a faster rate?

Sources and Suggested Readings

Line Graphs and Histograms
Wallgren, Anders, et al. *Graphing Statistics & Data*. Thousand Oaks, Calif.: Sage Publications, 1996.

ACKNOWLEDGMENTS

Special thanks to my editor Joy Ohm for her patience, perseverance, and expertise, and to Stephen Frenkel and Jim Biles for their thoughtful comments on the first draft. I would also like to thank the following people for their assistance: Greg Anderson, Deborah Che, Mary Beth Cunha, Doug Deur, Mona Domosh, George Erickcek, Anne Gibson, Susan Hanson, Mike Hass, Knud Larsen, Dave Lemberg, Bruce Macdonald, Michael McDonnell, Judy Olson, Tim Robinson, Stefan Sarenius, Jonathan Start, Hans Voss, Judy Walton, Kathleen Weessies, and everyone at W.H. Freeman.

These exercises were created from the resources (especially the map libraries) of Western Michigan University, Michigan State University, and Humboldt State University. I would also like to thank the Department of Geography at Western Michigan and the Department of Geography at Humboldt State for all of the encouragement and support.